CHALLENGER:
An American Tragedy

CHALLENGER:
An American Tragedy

The Inside Story
from Launch Control

HUGH HARRIS

INTEGRATED MEDIA

NEW YORK

All rights reserved, including without limitation the right to reproduce this book or any portion thereof in any form or by any means, whether electronic or mechanical, now known or hereinafter invented, without the express written permission of the publisher.

Copyright © 2014 by Hugh Harris

All photos courtesy of NASA

ISBN: 978-1-5040-7391-2

Published in 2022 by Open Road Integrated Media, Inc.
180 Maiden Lane
New York, NY 10038
www.openroadmedia.com

This book is dedicated to my wife, Cora; the Challenger *crew; and the family of tens of thousands of people who worked so hard to make the space shuttle program the success it was.*

INTRODUCTION

January 28, 1986, was one of the worst days of my life. This book explains why. This story, told by Hugh Harris, tells about the *Challenger* accident as only an insider could tell it. Hugh Harris was the public affairs commentator in launch control during the accident. He saw the whole event in vivid detail from the launch countdown to the tragic accident and its aftermath. He knew the people involved, including the crew, in great personal detail. The accident shook NASA and me personally to our very core.

Hugh was the voice of NASA to millions of people across the globe as he described the *Challenger* countdown from the Kennedy Space Center and its launch into a crystal-clear, but very cold, sky. Seventy-three seconds later *Challenger* was engulfed in a great explosion by the propellant that was supposed to power it into orbit.

Hugh has spoken personally to many who were on the job that day and found their lives forever changed by the experience.

One of the most important goals of the book is to emphasize, as he said, that "the real tragedy of an event like *Challenger* is in the loss of people and the accomplishments and inspiration they would have contributed to humankind."

I agree with that thought completely and would like to say a few words about the *Challenger* crew that were lost that day. All of the five members from the Astronaut Office were close personal friends of mine. I knew the two payload specialists very well, and they were both impressive individuals.

Commander Dick Scobee was the type of man who did not let things stand in his way, yet was one of the nicest people you'd ever want to know. Dick served as my pilot on my third flight. We became close friends. He began his air force career as an engine mechanic but used his time off to graduate from college. He went on to become a fighter pilot and an accomplished test pilot, and was selected as an astronaut twenty-three years after entering the air force. He never stopped learning or doing his best.

Pilot Mike Smith said, "I can never remember anything I wanted to do but fly." He was a guy who could do anything. One of his important accomplishments in the Astronaut Office was to devise a system that would allow crews to land the shuttle at night. He tested several concepts but finally settled on just lighting up the runway with large floodlights. That technique was used successfully on all shuttle night landings. Mike had already been assigned a second flight even before this one flew.

Ron McNair spent his childhood in the South, where he protested the library that tried to deny him, a nine-year-old African American boy, the privilege of borrowing books. Ron went on to achieve a doctorate in physics and to be selected as an astronaut. That same library is now named after him. He held a fifth-degree black belt in karate and was a performing jazz saxophonist.

INTRODUCTION

Ellison Onizuka was a great flight test engineer. He always had a smile on his face, which made everyone else smile as well. It is a quote from him that you find in American passports today: "Every generation has the obligation to free men's minds for a look at new worlds . . . to look out from a higher plateau than the last generation."

Judy Resnik was a poster woman for the fact that talents in math and science often go hand in hand with a talent for music. She moved effortlessly from being a classical pianist to electronic circuit designer to human researcher at the National Institutes of Health to astronaut. She was beautiful as well as brilliant.

Sharon "Christa" McAuliffe, the "Teacher in Space," was a fantastic ambassador for educators everywhere who instill the thirst for knowledge in our young people and, in so doing, shape the future of the human race.

Payload specialist Gregory Jarvis was another musician (he played classical guitar), engineer, and scientist. A communication satellite expert, he was deemed essential enough to follow the LEASAT from construction to its planned launch from the *Challenger*.

Some people have described the space station as being of the same magnitude of importance to the US as the pyramids were to the people of Egypt. They are wrong. Pyramids are the result of technical know-how and innovation being used for the dead. The space program is all about improving life for the living.

This book has all the facts, but more importantly, offers insight into the people. The people are what the space program is all about. They are the ones who make it happen.

Robert L. Crippen
Pilot of STS-1, the first space shuttle mission
Commander, STS-7, 41-C, 41-G

CHALLENGER:
An American Tragedy

CHAPTER ONE

A Look Back Twenty-eight Years

Challenger was a spacecraft designed to transport, protect, and nurture its seven-member crew as it transported them beyond the limits of our home planet's life-support system. There, they would conduct experiments to improve lives on Earth. Among its passengers was the first civilian crewmember, the "Teacher in Space" Sharon Christa McAuliffe (known as Christa), who was already inspiring a generation of school children.

I had watched from the firing room as the twenty-four previous shuttles rocketed upward and successfully returned to Earth. But on January 28, 1986, *Challenger* was engulfed in a fiery inferno in full view of thousands of people at the center and millions of others viewing the launch on television.

The tragedy produced a myriad of human emotions. For Todd Halvorson of *Florida Today*, it was an unforgettable introduction to space reporting. Hired the day before, but not yet on the job, he stepped out of the Cocoa Beach Holiday Inn to watch. Burned into his psyche are the pitchfork contrails and the memory of a weeping young girl, pointing upward and crying over and over, "The teacher is up there! The teacher is up there!"

For some young astronauts, it was a "loss of innocence" that took some time to accept. Franklin Chang-Díaz flew on STS-61C, the shuttle mission just a few weeks before *Challenger*. He and his crew experienced the tragedy from a viewing room at the Johnson Space Center in Houston.

"I think we were all unprepared to deal with this kind of event," he says. "From my first flight before the *Challenger* disaster, to my second flight, after, it felt as if we had lost our innocence. When I went into my second flight—well, it was probably the same way a soldier goes into battle with a few scars. You don't look at that battlefield the same way you did on the first day. I mean, it was still exciting, it was still wonderful, but we realized it was not child's play anymore."

Lisa Malone, then a young public information specialist who would become director of public affairs for the Kennedy Space Center (KSC) twenty years later, recalled, "At the time, I was angry. I was angry at the engineers. I didn't yet realize how hard space flight was. Later, as I started to go to more technical meetings, I learned the difficulty of managing risk posed by a highly complex vehicle."

The accident triggered in-depth investigations and denied the nation of human access to space for almost three years. Unmanned launches continued, but our astronauts stayed on the ground.

It brought into question the way management and technical experts worked together. It highlighted the role played by political decisions and uncertain year-to-year funding. It exposed the roadblocks to communication imposed by managers and organizational culture.

It was a chilling reminder that it is safer to sit on the ground than fly into space. But that's not an option for the human race.

Ultimately, it helped enable 110 more space-shuttle flights and the construction of the International Space Station, which ranks near the top of human achievement.

Dozens of people gave the "go" to launch on that morning twenty-eight years ago, and tens of thousands more had worked on the hardware. Yet, despite all of the investigative probing and some rancorous finger-pointing in the months to follow, no one ever alleged less than a strong desire to do his or her job to the best of his or her ability.

It demonstrated, once again, how much there is to learn as humankind continues to advance the boundaries of science, technology, and human interaction.

On that day, I was the chief of public information for NASA's Kennedy Space Center and the launch commentator. This piece will take you on the same journey I experienced in the hours before launch and then along the bumpy road to find the cause of the accident and heal the system.

CHAPTER TWO

A Cold, Cold Night

The night of January 28, 1986, was the coldest I can remember in Florida. But when I left my house in Cocoa Beach at two a.m., I wasn't thinking of the cold. I was worrying about getting to the Kennedy Space Center on time.

Every time I served as "the voice of launch control" for a space shuttle launch—a responsibility I had held beginning with STS-1 in 1981—I worried that my car would be delayed by the hundreds of thousands of people who came to watch. If I didn't get to the firing room on time, the launch would have happened anyway, but I would have felt like I'd let down the team.

But this morning, as I drove toward KSC, I did not find the usual congregation of cars. Very few were parked along the causeway over the Banana River. Normally, even at that early hour in the morning, and eight or more hours before a launch, the causeways were crowded. Families would leave their cars to make new friends or gather around radios to keep track of the progress of launch preparations. Cars would sport license plates from dozens of states—California, Washington, even Alaska. The space program was a source of national pride, and we who were privileged to work in it could not help but be inspired.

But this night was different. The few who had come were huddled inside their vehicles.

In the distance, Pad 39 B and *Challenger* were sparkling in the pure white light of the xenon searchlights. The thick shafts of light illuminated the rocket vehicle and slanted skyward for many miles.

As I drove toward the center along State Route 3 on Merritt Island, some of the orange groves huddled under blankets of smoke from large bonfires created to help protect the fruit from freezing. Most of the large groves had been flooded or sprayed with water. The temperature of fruit encased with ice does not drop below freezing. Smudge pots were no longer used due to pollution.

The air temperature was in the low thirties and dropping rapidly into the twenties. The smaller grove owners could not afford to protect their groves, and a week later their oranges would be thudding to the ground at the rate of a dozen per minute.

The officers at the first guard gate wore heavy jackets. "Do you think it will go, Mr. Harris?" one asked.

I told them the launch had already been postponed an hour and might be delayed further because of concern due to the cold. I said, "They're supposed to start tanking around three a.m. If they tank, they'll try to launch. They have about a two-hour window."

In capitulation to the freezing cold, the press site looked pretty deserted when I arrived. Normally, the photographers and reporters would be walking between buildings or gathered in little groups for a smoke. This morning they were all indoors.

There were fewer press representatives than normal, as well. The shuttle launches had become routine through the years. About five hundred media had been accredited for *Challenger*—as opposed to five times that number for STS-1. As I recall, only one of the major networks was covering the launch live.

The science writers typically on hand for launch had a conflict this time. Press briefings were taking place at the Jet Propulsion Laboratory in California, where many of the most knowledgeable space reporters were learning what scientists were discovering as *Voyager* flew past Uranus. Laurie Garrett of National Public Radio described the experience by saying, "Every single minute Uranus was blowing our minds more than the minute before. The moons of Uranus were absolutely the most stupendously puzzling things any of us had ever covered."

The twelve-acre press site is located at the Banana River Turn Basin, slightly more than three miles from the launch pads. During the Apollo program, barges bringing the rocket stages from Michoud Assembly Facility, just outside of New Orleans, unloaded at the turn basin; now it was the shuttles' external tanks that were unloaded there. It is just across the road from the Vehicle Assembly Building (VAB), where the Solid Rocket Boosters (SRB), external tanks, and orbiters were bolted together on a mobile launch platform before being taken to one of the two launch pads, designated Pads 39 A and 39 B. The *Challenger* launch was taking place from 39 B.

A three- to four-acre, six-foot-high mound had been built along the back of the press site with material dredged up to deepen the turn basin. On top was a 350-seat grandstand fitted with long counters, telephone hookups, and folding chairs, as well as several permanent structures put up by NASA, the major television networks, and the wire services. Another half-dozen office trailers had been brought in by *Florida Today*, the Orlando *Sentinel*, the Nikon camera company, and others were split between the mound and the lower level.

The public information office, my home away from home, was located at the press site in a geodesic dome originally bought for the United States Bicentennial Exposition. It also provided working space for media who didn't have their own facilities. KSC office spaces lined one inner wall of the dome; along the other were several rows of long, counter-like desks for the press with assigned spaces where they could order temporary phone hookups. There were bins for fact sheets and news releases and a bank of pay phones.

A waist-high counter separated the press from the information people and provided space for the press to ask questions. Members of the press were not allowed behind the counter unless they were invited in for an interview or other business.

The flags of the sixteen countries that were partners with the United States for the Spacelab missions and for the future International Space Station flew over the press area.

Down below the mound were several acres of grass and the large, iconic countdown clock at the water's edge. Many news photographers had used the countdown clock in the foreground of pictures of previous launches. Thousands more posed with it as proof that they had covered history.

For each launch, temporary grandstands were trucked in to accommodate about a thousand VIP visitors. These included the extended families of the astronauts who were flying and guests invited by NASA headquarters or other centers. The immediate families of the astronauts and special guests, such as members of Congress, would watch from the roof of the Launch Control Center.

Approximately twenty thousand other invited guests would be taken by bus or given car passes to park on the causeway across the Banana River connecting KSC and the Cape Canaveral Air Force Station about seven miles from the pads. Loudspeakers set up in each location allowed my commentary to keep them informed about what was happening. Public affairs representatives and car parkers at each location helped direct them and answer questions.

I reached the press site about eight hours before the then-scheduled launch time of 10:38 a.m. and went into my office after checking with the staff and saying hello to the press who had come in early. Almost everyone commented on the cold and speculated that we would postpone the launch for a third time.

The launch scheduled for two days earlier had been canceled because of the weather forecast. It turned out to be a perfect day. The attempt of the previous day, January 27, had been scrubbed because sensors showed that the crew ingress door on the *Challenger* was not securely closed. Once that was corrected, the handle used to latch the door could not be removed without drilling out the bolts. Time ran out, and crosswinds at the shuttle landing facility became unacceptable.

Finally we were at January 28. The day had everything going for it in terms of weather, except the bitter cold.

The first person I called was information specialist Andrea Shea King, who was in the firing room, keeping the press informed on the progress of loading liquid hydrogen and oxygen. "What are you hearing on the OIS?" I asked, referring to the Operational Intercom System, which tied all elements of the launch team together on more than thirty voice circuits.

"It's been pretty smooth, except for concern about ice on the pad," she reported. "The temperature is below thirty-two degrees. All the valves on the water lines on the pad have been open slightly all night so that they don't freeze. Can you see the icicles?"

Heavy ice coated the structures on the launch pad the morning of January 28, 1986.

The closed-circuit TV system showed an eerie winter scene. Long icicles hung from horizontal beams and cables on the fixed service structure, the tower housing the high-speed elevator that took the astronauts up to the 197-foot level. A swing-arm walkway allowed them to reach the White Room, where they entered the crew cabin. Another swing arm at the 215-foot level carried the "beanie cap" that fit over the top of the external tank. It collected and carried away the gaseous oxygen boiling off the liquid oxygen tank located in the top portion of the external tank.

The flickering flames from the hydrogen flare stack about a hundred yards from the shuttle added another surreal dimension to the scene at the pad. As the liquid oxygen, at 297 degrees below zero Fahrenheit, and liquid hydrogen, at 423 degrees below zero Fahrenheit, flowed through the piping and filled the tanks, the metal contracted and groaned, almost as if the vehicle were alive.

Very few people were allowed on the pad once the propellants started to fill the half-million gallon external tank. There was a close-out team made up of five people who performed the last-minute preparations for the astronaut crew and helped them strap into their seats. A second group was called the ice team. Their job was to look at the launch vehicle from every angle and make sure no ice was forming on the various connections and the surface of the tank. Ice chunks could indicate tiny leaks, or fall during launch to damage the orbiter's thermal-protection system. This could jeopardize a safe return into the atmosphere at the end of the mission. Since the ice team could not get close to every area of the vehicle, the team was equipped with infrared sensors so they could tell from a distance of hundreds of feet whether there were any unexpected warm or cold spots on the vehicle.

On this morning the team was expanded and brought in early because of the extreme cold. In addition to looking for ice build up from condensation, they were tasked with breaking off as many icicles as they could reach to make sure they would not fall on the vehicle at lift-off.

Liquid water is an important tool on every launch pad. Beginning at T-minus eleven seconds, three hundred thousand gallons of water would be sprayed over the mobile launch platform at the rate of a million gallons a minute to prevent sound energy from *Challenger*'s engines from being reflected back to the vehicle. Additionally, to absorb acoustic energy from the solid motors, dozens of long, water-filled canvas troughs were stretched across the openings in the launch platform underneath the solid rocket boosters.

Even though there was some antifreeze in the troughs, it was not enough. The water had turned to slush and then ice during the night. Its hard surface would have bounced the acoustic energy back toward the vehicle at the moment the solid motors ignited, possibly damaging the tail end of the orbiter. The ice team used fifty-foot-long poles to break up the ice. The troughs were filled again, this time with a stronger antifreeze mixture.

Seven miles to the south, things were about to get busy in the crew quarters. The pad workers would be able to go home soon after launch, but for the astronauts, it was the start of a busy day.

Nancy Gunter, who was in charge of the astronaut crew quarters in the operations and checkout building, had come to work before midnight. In addition to managing the team that kept the crew quarters at the ready, she was tasked with waking up the astronauts at the right time to get ready for their mission.

And though it wasn't in her job description, during her many years there, she baked most of the large ceremonial cakes that carried the crew patch in frosting on launch days. Despite the fact that every crew had breakfast at a different time of the day or night, depending on launch time, the cake was a centerpiece of the astronauts' breakfast table.

The astronauts never ate the cake before launch. In fact, some of them didn't eat anything before launch in order to ward off space sickness, which is very similar to seasickness. The cakes were frozen and often sent by plane to the Johnson Space Center in Houston to be consumed after the crew returned home.

As soon as I saw the TV feed of the crew—Commander Francis R. Scobee; Pilot Michael J. Smith; Mission Specialists Judith A. Resnik, Ellison S. Onizuka, and Ronald E. McNair; and Payload Specialists Gregory B. Jarvis and Christa McAuliffe—eating breakfast, I gathered up my countdown manual and reference material and got ready to go to the launch control center.

But first I called my boss, Charles T. Hollinshead, the director of public affairs for KSC. He held my job during the Apollo era and did the NASA commentary for the final Apollo launches.

I briefed him on the activities on the pad and made sure he had the phone number for my console in the firing room. He would be escorting Shirley Green, the new assistant administrator for public affairs, from NASA headquarters to the firing room later in the count. A veteran of public affairs activities in the White House, she had moved to NASA just a month or so before *Challenger*.

I assured him I would check out both of their consoles in the operational-support area of the firing room before they arrived.

CHAPTER THREE

The Launch

While the crew was finishing breakfast, I hopped in my government car for the short drive to the launch control center. Firing Room 3 was on the third floor. It was brightly lit and—thankfully—reasonably warm.

The firing rooms were set up with one thought in mind: communication.

They were arranged a little like amphitheaters. Seated on the highest level, immediately in front of the huge 1,600-square-foot windows looking out to the launch pads, were Launch Director Gene Thomas and other operational managers. I was the only one without direct supervision of the launch team present, and sat at the end of the row. I was also the only one to have a Plexiglas panel between my console and the one next to me. Since I would be broadcasting what was happening via NASA satellite to the news media at the press site and all of the viewing areas every few minutes, we tried to prevent the people next to me from being distracted.

On either side of the top row was a glass-encased triangular room. The mission management team occupied the left one, called the Operation Management Room (OMR). The NASA Administrator, the KSC director of public Affairs, the NASA assistant administrator for public affairs, and various other officials had consoles in the room on the right, called the Operation Support Room (OSR). Acting NASA Administrator Dr. William Graham would typically have been seated in the OSR, but was not present for launch, leaving General Manager Philip Culbertson as the senior NASA official present.

One row down, in the main room, were test conductors of the various shuttle elements. The director of safety and quality assurance and the flight surgeon were located another step down.

Each space shuttle had more than a million working parts, and keeping track of how everything was working was not easy. Situated on the huge floor of the firing room were semicircles of computer consoles devoted to the various elements of the launch process. The first group of consoles encountered upon entering through the security door at the lowest level was responsible for making sure all the communications, avionics, and navigations flight hardware was operating properly. Clear and intelligible voice and, frequently, video communication between the crew, the launch team, and flight team is essential during every aspect of a launch and mission.

The payloads systems console came next. Most of the payload bay of the *Challenger* that day was taken up with the thirty-two-thousand-pound tracking and data relay satellite and its second-stage motor. This was the second in a series of three satellites that would replace most of NASA's ground-based tracking stations around the world. It would also greatly increase the time during which mission control could monitor all the shuttle systems while it was on orbit and talk to the crew. The crew planned to deploy it just ten hours after reaching orbit. The other major payload was a low-cost, free-flying spacecraft named *Spartan*, which was carrying an array of instruments to study Halley's Comet.

While both payloads were basically "asleep" during launch, engineers at the payload console in the firing room and the spacecraft control center at NASA's Goddard Space Flight Center in Maryland would be monitoring its systems to make sure the temperatures, gas pressures, and electrical activity stayed the way they should.

Behind the payloads systems console were the engineers in charge of the environmental control and fuel cell systems. They monitored the devices that gave the crew the air they breathed, water they drank, and electrical power to run the orbiter systems. Water for cooling loops, sanitation, and cleaning the air was part of their responsibilities, as well as detection and purging of any vestige of hazardous gases on the pad or in the vehicle. Next to them, the propulsion team monitored the loading of half a million gallons of liquid oxygen and liquid hydrogen and the readiness of the engines to lift 4.5 million pounds of space hardware into Earth orbit.

Behind the environmental console were the electrical engineers who monitored the distribution of power from the launch pad and, late in the count, the electrical output of the fuel cells on board the orbiter. They probably had more onboard flight hardware and ground support equipment to maintain than any other flight systems engineers.

The environmental engineers shared their console space with the range safety engineers. The NASA contractor and the air force representatives at that console were able to stop the countdown up to the instant of main engine ignition at T-6.3 seconds. They were in constant contact with the Air Force Range Safety Center on Cape Canaveral and could destroy the shuttle if it were to go astray. This day would be the only time in the 135-flight program they would have to act.

In the back of the room was one of the most sensitive groups, the integration consoles. The consoles held the ground launch sequencer that continually monitored and managed more than two thousand measurements of everything from voltage and amps to fluid levels, gas pressures, temperatures, and valve positions. It had to ensure that more than five hundred measurements were correct during the last thirty-one seconds of the launch sequence alone.

I entered the firing room on the lower level after showing my badge to the guard at the door and climbed to the top level, waving quick greetings to people as I went by. Chief NASA Test Conductor Norm Carlson gave me a fifteen-second briefing on how things were going: "All the technical details are going well. We're still concerned about ice on the pad." Countdown was still at T-minus three hours and holding.

A countdown is an exacting process and sometimes a little confusing to the outside world. The word *countdown* refers to both the physical work performed by the many engineers and technicians involved as well as the six-volume set of procedures. The countdown manual specifies every task that needs to be completed, who does the work, the order in which it is done, and how long before launch it needs to be done. In the case of *Challenger*, the countdown started almost three days before launch. However, the countdown clock started at T-minus forty-three hours to allow extra time, called built-in holds, in case something went wrong or took longer than anticipated. The tasks ranged from pumping the propellants on board to aligning the guidance system.

Each launch time is determined by calculating where a spacecraft is going in the universe; whether Earth orbit is desired; how much sun is needed on the solar panels; or the trajectory and energy needed to rendezvous with another spacecraft. In *Challenger*'s case, it was based largely on the position needed to launch its payload satellites into the proper geosynchronous transfer orbit.

Each time the launch of *Challenger* was delayed, the orbital mechanics experts in mission control at the Johnson Space Center in Houston had to recalculate the launch time based on all of those considerations. Every launch involves other considerations and rules, such as the amount of time the astronauts can be subjected to lying on their seats in Earth gravity. Five hours is considered the limit of "crew on back" time. Once they're in zero gravity, of course, any position is equally comfortable.

In the Atlantic Ocean east of Cape Canaveral on that blustery January morning, things were not comfortable aboard the solid rocket booster recovery ships, *Liberty Star* and *Freedom Star*. After launch, the boosters would parachute into the Atlantic Ocean approximately 140 miles east of the cape. The seas in that area had thirty-six-foot waves and winds of about forty knots, gusting up to sixty knots. As the seas kicked up, the ships had to seek calmer waters closer to shore. This meant that NASA might not be able to recover some of the normal hardware ranging from the parachutes and the nose portion of the SRBs to the solid motors themselves. Reuse of the solid motors and other hardware saved millions of dollars each flight.

Because there was some flexibility in the *Challenger* launch, T-minus zero—launch time—had been postponed one hour during the night, allowing the crew an extra hour of sleep before they were awakened. Launch was now set for 10:38 a.m. (EST).

The *Challenger* crew in the White Room just before launch: Payload Specialists Sharon Christa McAuliffe and Gregory B. Jarvis; Mission Specialist Judith A. Resnik; Commander Francis R. Scobee; Mission Specialist Ronald E. McNair; Pilot Michael J. Smith; and Mission Specialist Ellison S. Onizuka.

At 7:48 a.m., the crew finally left the operations and checkout building to be driven to the pad. Their entry into the orbiter went smoothly this time; the hatch closed and the handle that had caused the previous day's postponement was removed without incident.

The launch time was postponed another hour during a hold at T-minus one hour, ten minutes, to send the ice team back to the pad to reassess the situation and remove more icicles from the fixed service structure and the water troughs. Launch was now set for 11:38 a.m.

Delays are almost routine for the crew and launch team. In the orbiter, the crew was relaxed and jovial. Everyone wanted conditions to be perfect and was willing to wait until every problem was resolved. On this day, many members of the launch team told me later, they thought we would scrub and come back another day.

But finally we were underway. In the firing room, I had taken over broadcasting the events at about 7:45 a.m., keeping the press and world informed about the milestones leading up to launch. We came out of the final built-in hold at T-minus nine minutes, and the ground launch sequencer took command of the countdown.

I continued broadcasting. "T-minus seven minutes, thirty seconds and counting. The walkway the astronauts used to enter the *Challenger* is being swung away from the orbiter. It can be put back in place in twenty seconds in case of an emergency."

Inside the *Challenger*, the crew was excited. Joking among themselves to relieve everyone's tension is a time-honored tradition by space fliers, but doesn't get broadcast to the rest of the world.

While Pilot Mike Smith was flipping the proper switches, I was telling the press and guests, "T-minus five minutes. Auxiliary power unit start." The hydraulic power units were used to control the various aero-surfaces and position the orbiter's main engines. They were all moved through preprogrammed maneuvers to make sure they were ready to steer the vehicle properly during launch and landing.

At T-minus four minutes the crew was instructed to close their visors, sealing their helmets.

As the count approached T-minus two minutes, Mission Specialist Judy Resnik teasingly asked for her security blanket back.

Commander Scobee told Payload specialists Christa McAuliffe and Gregory Jarvis, who were seated below the flight deck, "Two minutes downstairs," and asked, "You got a watch running down there?"

At T-minus one minute, forty-seven seconds, Pilot Smith announced: "Okay, there goes the LOX arm." The liquid oxygen (LOX) arm allowed the oxidizer to flow into the external tank.

At T-minus one minute, forty-six seconds, Scobee said, "[There] goes the beanie cap."

T-minus one minute, forty-four seconds, Mission Specialist Ellison Onizuka quipped, "Doesn't it go the other way?"

T-minus one minute, forty-two seconds. Everyone laughed.

T-minus one minute, thirty-three seconds. Resnik reminded people, "Got your harnesses locked?"

T-minus one minute, twenty-nine seconds. Smith said, "What for?"

T-minus one minute, twenty-eight seconds. Scobee: "I won't lock mine; I might have to reach something."

T-minus one minute, twenty-four seconds. Smith didn't believe that and said, "Ooh-kaaaay."

T-minus fifty-nine seconds. Scobee, "One minute downstairs."

T-minus fifty-two seconds. Resnik reminded everyone, "Cabin pressure is probably going to give us an alarm."

Simultaneously, I was reminding the media and launch crowd: "Caution and warning system alarms routinely sound at this time, indicating the cabin pressure is acceptable . . ."

T-minus fifty seconds. Scobee: "Okay."

T-minus forty-three seconds. Smith: "Alarm looks good."

T-minus forty-two seconds. Scobee: "Okay."

T-minus forty seconds. Smith: "Ullage pressures are up." (Ullage is the empty space in the tanks, which increases in pressure as the propellants continue to boil off from liquid to gas after the exhaust valves are closed.)

T-minus thirty-four seconds. Smith: "Right engine helium tank is just a little bit low." (Helium pressure to the main engines.)

T-minus thirty-two seconds. Scobee: "It was yesterday, too."

T-minus thirty-one seconds. Smith: "Okay."

T-minus thirty seconds. Scobee, "Thirty seconds down there."

T-minus twenty-five seconds. Smith, "Remember the red button when you make a roll call." (Precautionary reminder for communications configuration.)

T-minus twenty-three seconds. Smith, "I won't do that; thanks a lot."

T-minus fifteen seconds. "Fifteen seconds and counting," I say, and continue counting for the outside world: "Ten, nine, eight, seven, main engine start."

T-minus six seconds. Scobee told the crew, "There they go, guys."

T-minus five seconds. Resnik, "All right."

T-minus three seconds. Scobee, "Three at a hundred." (Main engines at 100 percent thrust level.)

T-minus zero. I completed my commentary by saying, "Lift-off, lift-off of STS-51-L, and it's cleared the tower." At tower clear, my job was over. Commentary switched to Steve Nesbitt in mission control in Houston.

T-plus zero. Resnik, "Aaall riiight."

T-plus one second. Smith, "Here we go."

T-plus seven seconds. Scobee, "Houston, *Challenger* roll program."

Steve Nesbitt announced, "Roll program confirmed. *Challenger* now heading downrange . . ."

T-plus eleven seconds. Smith, "Go, you mother."

T-plus fourteen seconds. Onizuka, "LVLH." (He was reminding the crew of the cockpit switch configuration change [local vertical/local horizontal].)

T-plus fifteen seconds. Resnik, "[Expletive] hot."

T-plus sixteen seconds. Scobee, "Ooohh-kaaay."

T-plus nineteen seconds. Smith, "Looks like we've got a lotta wind here today." (*Challenger* is being buffeted by the wind shear.)

T-plus twenty seconds. Scobee, "Yeah."

T-plus twenty-two seconds. Scobee, "It's a little hard to see out my window here."

T-plus twenty-eight seconds. Smith, "There's ten thousand feet and Mach point five." (The altitude is ten thousand feet, and the speed is half the speed of sound.)

T-plus thirty seconds. (Garbled sound.)

T-plus thirty-five seconds. Scobee, "Point nine." (Nine tenths the speed of sound.)

T-plus forty seconds. Smith, "There's Mach one."

T-plus forty-one seconds Scobee, "Going through nineteen thousand [feet]."

Nesbitt in Houston: "Engines getting throttled down now. At 94 percent. Normal throttle for most of the flight is 104 percent. We'll throttle down to 65 percent shortly. Engines at 65 percent. Three engines running normally. Three good fuel cells. Three good APU's. Velocity 2,200 feet per second."

T-plus forty-three seconds Scobee, "Okay, we're throttling down."

T-plus fifty-seven seconds Scobee, "Throttling up." (Engines are throttled back up to 104 percent after going through the region of maximum dynamic pressure in the atmosphere.)

T-plus fifty-eight seconds Smith, "Throttle up."

T-plus fifty-nine seconds Scobee, "Roger."

T-plus sixty seconds Smith, "Feel that mother go."

T-plus sixty seconds "Woooohoooo."

T-plus one minute, two seconds Smith, "Thirty-five thousand going through one point five." (The altitude and velocity report: thirty-five thousand feet, Mach 1.5.)

T-plus minute, five seconds Scobee, "Reading four eighty-six on mine." (Routine airspeed indicator check.)

T-plus minute, seven seconds Smith, "Yep, that's what I've got, too."

T-plus minute, ten seconds Scobee, "Roger, go at throttle up."

T-plus minute, thirteen seconds Smith, "Uh-oh."

T-plus one minute, thirteen seconds. (LOSS OF ALL DATA.)

Nesbitt, watching the KSC television feed and instrument reports in mission control, said, "Velocity 2,900 feet per second. Altitude: nine nautical miles. Distance downrange: seven nautical miles."

Either Resnik or Onizuka, or both, reached down to activate their Personal Egress Air Packs as well as the one for Smith, as they had trained for in emergency drills.

As we watched the shuttle on the monitors in the firing room, it looked as though the flames from the solid motors were beginning to creep upward to consume the entire orbiter. It took only a blink of the eye, but the moment seemed to last a lifetime. Suddenly, *Challenger* was engulfed by a huge fireball.

From Houston, Nesbitt announced, "Flight controllers here looking very carefully at the situation. Obviously, a major malfunction. We have no downlink."

Shuttle Operations Director Bob Sieck said to no one in particular, "It's a bad day!"

The two solid rocket motors spiraled out of the flames, creating a pitchfork-shaped contrail.

The iconic "forked contrail" seen immediately after the accident as the two solid rocket motors spiraled away from the orbiter.

"A very bad day," Sieck muttered.

Launch Director Gene Thomas bowed his head in silent prayer.

CHAPTER FOUR

After the Launch

It was very quiet in the firing room. For a brief period, it seemed that everyone held a collective breath.

I, and probably most other people there, had seen large unmanned rockets explode during flight. We had seen the twisted metal remains that were recovered. I don't think anyone really believed the astronauts had survived. Nevertheless, we all watched the tracking camera feeds in hopes of seeing the *Challenger* orbiter emerge, intact, from the ball of flames that covered several square miles of sky.

It didn't happen. Later analysis of the footage revealed that the crew cabin was ripped from the orbiter, continued upward to sixty-five thousand feet, then began its long fall downward. It impacted the ocean minutes later.

Then, after a stunned silence and despite many wet eyes, professionalism took over.

The team began emergency procedures. The pad was returned to a safe condition. An announcement was made on the public-address system that anyone leaving the firing room had to sign out and could not take any papers or notes they had made during the launch process. All documentation was to be impounded in case it might shed light on the cause of the accident.

In a slightly shaky voice, Marvin Jones, who had recently joined NASA, after serving as the air force commander of the Eastern Test Range, called his successor and told him what we were seeing and to expect requests for air force helicopters and other assistance.

Center Director Dick Smith turned to Launch Director Gene Thomas and said, "What happened?" Smith later recalled his question as "probably the most stupid statement I've ever made in my life. It was obvious what happened. You just were not willing to accept it."

Actually, it was exactly the question that everyone was asking.

I picked up the phone and called Gatha Cottee, one of our KSC television producers. "Copy all of the videotapes right away. Security will be coming to impound the originals for the investigation." In typical Cottee fashion, he overreacted and locked all the doors to the television building so the guards couldn't get in.

He also forgot that guards have master keys. I needn't have worried, though. Fletcher Hildreth, our RCA Contractor Television Director, and Producer Bill Johnson had it under control. Tapes were already being duplicated and both of them were calmly orchestrating the television coverage. Hildreth deftly kept the NASA public-affairs cameras from showing a parachute that deployed. He knew it would only confuse any viewers; the only parachutes on board were for the nose cones on the booster rockets. The *Challenger* was the first American spacecraft to carry astronauts but have no escape system.

Meanwhile, one of the volunteer helpers ran along the front of the press site grandstand yelling, "RTLS! RTLS!" *RTLS* stands for "return to launch site abort." All of us in the firing room knew that would not happen. An RTLS required both the solid motors to stay attached throughout their burn, and the main engines, fueled by the hydrogen and oxygen in the external tank, to slow the orbiter down. Instead, those propellants had created the fireball.

On top of the vehicle-assembly building, the photo escorts were told by Ed Harrison, the NASA KSC photo chief, via walkie-talkie: "We don't know what is going to happen; move the photographers to the northwest corner. If the shuttle is able to come back, that will be the best shot."

On top of the launch control center, where the immediate families of the astronaut crew watched in horror, Nancy Gunter helped form a line with the guards to make sure the other guests did not rush over to the families. The families were escorted off the roof by the public affairs escorts, Libby Wells and Bob Harris, and Astronaut Escort Frank Culbertson. There are always several people assigned to stay with astronaut family members during a launch to guide them to various locations and answer their questions.

This time, no one had any answers.

The families boarded their waiting bus and were whisked back to astronaut crew quarters to await new information. Several were taken to the Patrick Air Force Base hospital, as a precaution, to be treated for shock.

NASA's general manager, Philip Culbertson, was also watching from the roof, along with his daughter, Camden, and his two grandsons. As soon as he saw the fireball, he turned to them and said, "You'll have to find your own way home." Luckily his secretary was there to escort them back to their car.

At the VIP site, adjacent to the press site, several public affairs officers grabbed hands to say the Lord's Prayer—then ran to the press site to help answer the storm of incoming phone calls.

Richard Nelson, a senior design engineer, was at the VIP site that morning. "When my fellow engineers and I saw what was happening we sat down and started worrying that we had missed something. Our job was ensuring that we could detect and prevent any hazardous gases from building up in the vehicle during processing and launch. This included hydrogen that could cause an explosion or fire."

Nelson said, "Everyone was cheering and shouting as *Challenger* lifted off. The cheering went on even after the vehicle started to break up. We knew the families who were still cheering didn't realize what was happening."

According to Nelson, as the shouting ceased and the horror of the scene started to sink in "Steve Dutchak and some of the other public affairs escorts stepped out in front of the stands. In a kindly but professional way, almost like it was rehearsed, he told them to take their time, but when they were ready, the buses would take them back to their cars or offices.

"When I came down from the stands," Nelson says, "I had to walk in front of the bleachers where the family members were. There was a little girl standing by the white picket fence crying. Apparently she was one of Christa's students. There was a photographer with a camera that had a huge lens just taking picture after picture of her. I walked in front of the photographer, put my hand over his lens, and said, 'Stop that!' He said, 'I'm only doing my job . . .' and I told him, 'With that lens you can do it from a lot farther away' . . ."

Back in the firing room, I hurried across the top row to the operation support room where Chuck Hollinshead and Shirley Green were sitting. We discussed the long-established contingency plan, which called for NASA commentary to switch back to KSC. We would then report to the media and guests on what steps were being taken by the launch team and emergency personnel. I was told not to resume commentary.

This was the first decision that contradicted the contingency plan—and a bad one. The plan prescribed that we get a top NASA official to the press site within an hour to tell the media what was happening. When it actually took five hours to hold a press conference, it meant the flow of information stopped, and NASA looked like it was hiding from the world.

Engineers and scientists hate to talk to anyone, let alone the press, when they don't know all the facts. Associate Administrator of Space Flight Jesse Moore and the other managers spent that five-hour period reviewing film and telemetry information and organizing the investigation. At the end of the five hours, they still didn't know all the facts. But the delay was the beginning of a great loss of credibility for the agency.

Steve Nesbitt, the commentator in Houston, continued calmly for as long as he could. Twenty-one minutes after the accident, his final comments were:

"This is mission control, Houston. Repeating the information that we have at this time. We had an apparently nominal lift-off this morning at 11:38 eastern standard time. The ascent phase appeared normal through approximately the completion of the roll program and throttle down and engine throttle back to 104 percent. At that point, we had an apparent explosion. Subsequent to that, the tracking crews reported to the flight dynamics officer that the vehicle appeared to have exploded and that we had an impact in the water down range at a location approximately 28.64 degrees north, 80.28 degrees west.

"At that time, the data was lost with the vehicle. According to a poll by the flight director, Jay Greene, of the positions here in mission control, there were no anomalous indications, no indications of problems with engines or with the SRBs or with any of the other systems at that moment through the point at which we lost data. Again, this is preliminary information. It is all that we have at the moment and we will keep you advised as other information becomes available. . . . This is mission control, Houston."

Back in Florida, other contingency activities were going well.

Within two minutes of the accident, the landing and recovery director and the booster recovery director notified the SRB recovery ships, *Liberty Star* and *Freedom Star*, of the approximate latitude and longitude of the impact area. The two ships had been battling the Atlantic's waves to stay as close as possible to the normal SRB recovery area. Because of the clouds, those onboard had not seen the accident.

Now the ships turned eastward and increased their speeds to fifteen knots. As they went, they started finding and recovering pieces of floating debris.

An H-1 helicopter was notified to take off but hold short of the danger zone. Debris was estimated to continue falling for approximately fifty-five minutes.

The entire space-center phone system went out shortly afterward because of the volume of people calling in and out. What we didn't know immediately was that the telephone system for the entire area, including parts of Orlando, had simply overloaded and crashed. Restoring normal phone service took almost five hours.

Luckily, the intercom system continued to function, along with the walkie-talkie system the press site used. The combination was enough to coordinate the essential tasks.

All across the space center and throughout the surrounding community, people were coming together to express their collective sorrow. At the press site, one well-known broadcaster walked slowly into the office of Diana Boles, who was in charge of logistics and maintenance of the press site. She had been struggling with finding workers to fix the broken heating system, and the press dome was freezing that morning. Without a word, he gave her a hug and walked back out to do his job. It would not have been noteworthy except that their interactions were often contentious. But, for that moment, they were conscious of a shared sense of loss.

Makeshift memorial set up near Launch Pad 39 B by pad workers and guards shortly after the accident.

***

Back in the launch control center, Chuck Hollinshead and I were trying to get Associate Administrator of Space Flight Jesse Moore to commit to a press conference within the next hour. At the same time, Phil Culbertson had asked him to chair a meeting of senior managers associated with the launch to begin the process of organizing the investigation.

To keep the importance of a press conference in Moore's mind, we followed him down to the television control room to watch as tapes were played and replayed. Nothing definite that could have caused the accident popped out of the images immediately.

Press conferences were secondary in most of the managers' minds, but having one soon was uppermost in mine. Luckily, my office had the best news and audiovisual chiefs in the agency. Dick Young had been a newspaper reporter for years before joining NASA and was expert at working with the reporters. Ed Harrison started his career at Langley Research Center, where he photographed everything from work in wind tunnels to the earliest astronauts during their emergency training. Because they were handling things, I had the luxury of single-mindedly pursuing Jesse Moore for a press conference.

It took hours for Moore to work through the process of appointing the primary panel of experts. The center directors of the Kennedy Space Center and Marshall Space Flight Center, Shuttle Program Manager Arnold Aldrich, KSC Spacelab Director James Harrington, and NASA Consultant to the Administrator Walt Williams were primary members. Next came the development of a long list of action items.

Five hours passed from the time of the accident to the beginning of the press conference. The press was not happy. The delay allowed more media people to arrive, and there would have been even more if the phones had not been out. Within twenty-four hours the press corps would grow by another thousand.

There wasn't room enough for the media to crowd into the normal indoor space, so we quickly set up a table and microphones outside in front of the 350-seat grandstand. I introduced Jesse Moore, and he did the best he could with the sparse information we had at that point, telling the group:

"It is with deep, heartfelt sorrow that I address you this afternoon.

"At 11:40 this morning, the space program experienced a national tragedy with the explosion of the space shuttle *Challenger* approximately a minute and a half after its launch from here at the Kennedy Space Center.

"I regret that I have to report, that based on very preliminary searches of the ocean where the *Challenger* impacted this morning, that these searches have not revealed any evidence that the crew of *Challenger* survived."

During a brief question-and-answer period he announced that he was suspending shuttle operations. He added, "We will not pick up any flight activity until we understand what the circumstances were."

CHAPTER FIVE

Challenger and the White House

The president's State of the Union address is perhaps the most important speech he makes each year. Launch morning found President Reagan in debate with congressional leaders including House Speaker Thomas "Tip" O'Neill. Unemployment was the year's most controversial topic, and Reagan planned to make the subject a major part of his address that night.

As soon as the meeting with Congress ended, he began preparing for his State of the Union luncheon with news anchors from every television network. He was interrupted by his chief of staff, Donald Regan, and Vice President Bush telling him that *Challenger* had exploded.

After viewing television coverage of the accident, Reagan postponed the State of the Union Address and began work on what he would tell the nation later that day. He attended his scheduled meeting with the news anchors and other meetings, but by 5 p.m. was ready to address the nation.

President Reagan knew instinctively that his words would play an important part in the future of the country and the space program. He realized that one of his most important jobs was to provide comfort in time of crisis, to give assurance about the future and the leadership to move forward, and he stepped up to the task.

Vice President George H. W. Bush was close behind him.

The vice president made arrangements to fly down to Florida with Senators John Glenn and Jake Garn to comfort the families and speak to the NASA and contractor launch teams. Glenn was well known to the families, while Senator Garn had been the first non-astronaut to fly in space. He rode the shuttle the previous year, while serving as head of the Senate appropriations subcommittee, which dealt with the NASA budget.

Acting NASA Administrator Graham, who had been in Washington during the launch, arranged to fly back down with the vice president.

White House visits typically take place only after weeks of planning and on-the-ground surveys by the Secret Service. But by 5:30 p.m., Air Force One touched down at the Cape. Bush, Glenn, Garn, and Graham were whisked to the astronaut crew quarters to meet with the families. There, in an informal, family-type setting, Bush expressed the nation's sorrow and appreciation of the many contributions the crew had made to the country. He pledged that the cause of the tragedy would be found and fixed. Finally, he emphasized the importance of carrying on the space program. He said it was the only fitting tribute to the crew and the families.

In Firing Room 3, the vice president addressed the launch team. As he spoke of the crew, his eyes filled with tears. Launch Director Gene Thomas recalled, "I will always remember how this great leader actually shared in our sorrow by coming close to us where we worked."

The vice president went on to stress his and the president's faith in the NASA team and their ability to recover from the accident and continue an even stronger program.

Before flying back to Washington, Bush met with the mission management team. He was briefed on the steps already being taken to investigate, to find and fix the problems brought to light by the accident.

It is likely that more people in the country, and certainly in the area, listened to President Reagan's remarks on *Challenger* that same evening than would have tuned in to the State of the Union address.

It was inspirational to all of us involved.

The president said, "Ladies and Gentlemen, I'd planned to speak to you tonight to report on the State of the Union, but the events of earlier today have led me to change those plans. Today is a day for mourning and remembering. Nancy and I are pained to the core by the tragedy of the shuttle *Challenger*. We know we share this pain with all of the people of our country. This is truly a national loss.

"Nineteen years ago, almost to the day, we lost three astronauts in a terrible accident on the ground. But we've never lost an astronaut in flight; we've never had a tragedy like this. And perhaps we've forgotten the courage it took for the crew of the shuttle; but they, the *Challenger* Seven, were aware of the dangers, but overcame them and did their jobs brilliantly. We mourn seven heroes: Michael Smith, Dick Scobee, Judith Resnik, Ronald McNair, Ellison Onizuka, Gregory Jarvis, and Christa McAuliffe. We mourn their loss as a nation together.

"For the families of the seven, we cannot bear, as you do, the full impact of this tragedy. But we feel the loss, and we're thinking about you so very much. Your loved ones were daring and brave, and they had that special grace, that special spirit that says, 'Give me a challenge and I'll meet it with joy.' They had a hunger to explore the universe and discover its truths. They wished to serve, and they did. They served all of us.

"We've grown used to wonders in this century. It's hard to dazzle us. But for twenty-five years the United States space program has been doing just that. We've grown used to the idea of space, and perhaps we forget that we've only just begun. We're still pioneers. They, the members of the *Challenger* crew, were pioneers.

"And I want to say something to the schoolchildren of America who were watching the live coverage of the shuttle's takeoff. I know it is hard to understand, but sometimes painful things like this happen. It's all part of the process of exploration and discovery. It's all part of taking a chance and expanding man's horizons. The future doesn't belong to the fainthearted; it belongs to the brave. The *Challenger* crew was pulling us into the future, and we'll continue to follow them.

"I've always had great faith in and respect for our space program, and what happened today does nothing to diminish it. We don't hide our space program. We don't keep secrets and cover things up. We do it all up front and in public. That's the way freedom is, and we wouldn't change it for a minute. We'll continue our quest in space. There will be more shuttle flights and more shuttle crews and, yes, more volunteers, more civilians, more teachers in space. Nothing ends here; our hopes and our journeys continue. I want to add that I wish I could talk to every man and woman who works for NASA or who worked on this mission and tell them: 'Your dedication and professionalism have moved and impressed us for decades. And we know of your anguish. We share it.'

"There's a coincidence today. On this day 390 years ago, the great explorer Sir Francis Drake died aboard ship off the coast of Panama. In his lifetime, the great frontiers were the oceans, and a historian later said, 'He lived by the sea, died on it, and was buried in it.' Well, today we can say of the *Challenger* crew: Their dedication was, like Drake's, complete.

"The crew of the space shuttle *Challenger* honored us by the manner in which they lived their lives. We will never forget them, nor the last time we saw them, this morning, as they prepared for the journey and waved good-bye and 'slipped the surly bonds of Earth' to 'touch the face of God.'"

CHAPTER SIX

Reporters, Reporters Everywhere

It was very quiet in the press site dome while President Reagan was speaking. Even the phone-call volume dropped. However, it picked up again as soon as he finished.

The press site received 6,500 calls between five p.m. on the day of the accident and midnight on the first of February. Another six hundred calls couldn't get through because the lines were jammed. That's an average of more than two calls a minute, around the clock. It might have been more if the phone system hadn't failed.

Calls to the press site were supposed to be from journalists only, but we encouraged the operators to route calls to us if they didn't know how to handle them.

Many of the callers just wanted to express their sympathy about the astronauts and urge us to get back to flying safely as soon as possible. There were psychics who were sure they knew the cause of the disaster. Sabotage, which was ruled out very early in the investigation, was a favorite assertion.

Like many people who worked at the press site, Lisa Fowler, who was in charge of press accreditation for KSC, had some long days coming. She came to work at five a.m. launch morning and stayed until midnight, then was back at four a.m. the next morning. Following the launch, she badged 554 new reporters, more than had been badged in the weeks leading up to the launch.

The pressure on our public affairs personnel was extreme, but it was even greater on some of the individual reporters. Within forty-eight hours, there was a thundering herd of a couple thousand newspeople at the Kennedy Space Center, financed to the tune of millions of dollars by their organizations. Hundreds more descended on the Johnson Space Center, Marshall Space Flight Center, and other NASA centers. Their editors wanted good value for their money. They wanted the answers now, and they wanted them exclusively.

Worse, many of the reporters had never been to KSC or written about the space program before. But having reporters who needed a lot of help was not the biggest problem faced by press site personnel.

In normal times, we could call on engineers from every discipline to talk to reporters or explain how something worked. All of the experts were now assigned to various parts of the investigation.

NASA headquarters public affairs decided to make all substantive questions not asked at press briefings go through the Freedom of Information process. This process, usually reserved for legal or policy questions, slowed down our ability to supply answers.

Another decision that stirred up our working relationship with the media involved photography. During launches, photographers from the wire services, newspapers, and magazines are allowed to place cameras much closer to the pads than people are allowed at launch time. These cameras are automatically triggered by the sound or light as the rocket engines ignite.

Investigation managers decided that one or more of the media cameras that had recorded the launch might reveal the cause of the accident and, therefore, NASA needed to process the film and examine the images before the press could have them. It was not intended to prevent the news organizations from using the pictures, only to ensure they were preserved for the investigation.

The *New York Times* sued NASA.

Ed Harrison, our photography chief, worked tirelessly for several days to catalog, follow the processing, and return the pictures to the right media organizations.

Taken together, these decisions created a contentious relationship between the media and NASA, replacing the professional and harmonious atmosphere that prevailed before the accident.

CHAPTER SEVEN

The Commission

Politics also played a major part in the investigation.

NASA had taken the lead in past accident investigations, including the Apollo fire that killed three astronauts in a ground test. The agency started down that path immediately. But, this time, the White House decided to appoint a special outside investigative commission.

Four days after the accident, on February 3, President Reagan chose a stellar group of people from various walks of life. He selected former Secretary of State and former Attorney General William P. Rogers as chairman, with Neil Armstrong, former astronaut and then chairman of the board of Computing Technologies for Aviation, as vice chairman.

Other Commission members included General Chuck Yeager, first human to break the sound barrier; Dr. Sally K. Ride, first American woman in space; Robert W. Rummel, former vice president of Trans World Airlines, a space expert and aerospace engineer; Dr. Arthur B. C. Walker Jr., astronomer, professor of applied physics, and former associate dean of the graduate division at Stanford University; Nobel Prize Laureate Richard B. Feynman, professor of theoretical physics at the California Institute of Technology; Eugene E. Covert, then professor and head, department of aeronautics and astronautics at Massachusetts Institute of Technology; Robert B. Hotz, editor-in-chief of *Aviation Week & Space Technology* magazine from 1953 to 1980; David C. Acheson, former senior vice president and general counsel, Communications Satellite Corporation; Major General Donald J. Kutyna, USAF director of space systems and command, control, communications; Dr. Arthur B. C. Walker Jr., astronomer and professor of applied physics at Stanford University; and Dr. Albert D. Wheelon, physicist and executive vice president, Hughes Aircraft Company.

Shortly afterward, Joseph Sutter, executive vice president of the Boeing Commercial Airplane Company was added. Dr. Alton G. Keel Jr. was detailed to the Commission from his position in the executive office of the president, office of management and budget, as executive director.

Two days later, Dr. Graham renamed the original NASA investigation group the STS 51-L Data and Design Analysis Task Force. Jesse Moore continued as chairman until late February, when the Presidential Commission suggested that anyone directly responsible for the launch decision be removed from the investigation. Pending the arrival of a new chairman, NASA Astronaut Robert L. Crippen was named vice chairman and acting chairman.

At the time of the accident, Crippen had been preparing to command the first shuttle to be launched from Vandenberg Air Force Base in California. The launch had been tentatively scheduled for midsummer. Crippen, who had piloted STS-1 and commanded STS-7, STS-41-C, and STS-41-G, was at Los Alamos National Laboratory in New Mexico with his crew at the time of the accident. They were training on a classified Department of Defense payload planned for their flight.

Crippen had arranged to view the *Challenger* launch on TV in a conference room. He recalls, "At the time of the launch we were all watching it. Then about fifteen seconds after launch the network cut away to something else. We were walking out of the room when the network came back on and showed the shuttle coming apart.

"It was obvious to all of us that we had lost the crew and the vehicle . . . so our first thought was, we need to get back to Houston. We got a helicopter to take us back to Kirtland Air Force Base in Albuquerque, New Mexico. There the entire crew piled into our T-38s and flew back to Ellington, just outside of the Johnson Space Center.

"We reported to George Abbey, who was director of flight crew operation. He had flown back to JSC as soon as possible after the accident. He said he wanted me to go to the Cape and be the deputy to whoever would lead the investigation for NASA. In particular he wanted me to concentrate primarily on what had happened to the crew, along with the cause of the accident."

Another key player in the investigation was Jay Honeycutt, who would later become director of the Kennedy Space Center. Honeycutt was in the firing room for the launch, seated in the OMR beside Abbey. His primary responsibility for the launch was communicating with Astronaut John Young, who was flying the weather aircraft on the mission and gathering information for the mission management team from other sources. Although Honeycutt was the manager of operations for the space shuttle program office at JSC, he would soon find himself with many other responsibilities as part of the NASA investigation group.

In late February, former Astronaut Richard Truly became the new associate administrator of space flight. Truly had returned to the U.S. Navy after flying twice on the shuttle and was now an admiral. He left his job as commander of the naval space command to return to NASA and would eventually become its administrator.

Ironically, Truly's flight on STS-2 was the first time that hot solid motor gases penetrated the primary O-ring. Luckily, the secondary O-ring held and the mission was a success.

Truly named James R. Thompson chairman of the NASA investigation group. Thompson had recently retired from NASA and was a deputy director for technical operations at Princeton University's Plasma Physics Laboratory.

The agency was very respectful of the privacy and sensitivities of the astronauts' families. As a result, all news releases, photos, and answers to media queries relating to the families had to be routed through Bob Crippen or Jay Honeycutt.

CHAPTER EIGHT

SCOOP!!!

The press had three primary questions they wanted answered immediately. What caused the accident? Whose fault was it? Did the astronauts die instantly?

NASA wanted to know the answers, too.

Reportorial skill would quickly answer the first question.

Jay Barbree, who recently celebrated fifty-five years with NBC News, immediately started reaching out to his many contacts in the space industry. He was not going to wait for a NASA press conference to tell him what he wanted to know. Because he had lived in the area many years, and had made friends with everyone from astronauts to engineers to janitors, he was prepared.

A good friend of Barbree's and of mine, engineer and Space Shuttle Manager Sam Beddingfield, had just retired. Barbree called him. "Sam," he said, "I imagine you might want to visit some of your friends in the headquarters building and on the fourth floor and listen to what they're saying about the accident."

"I still have a badge. I think I'll do that," Sam agreed. He was as curious as anyone about what happened, and as devastated.

The next day, January 30, Sam called Jay. "I've got it," he said. The info was electrifying. "Sam!" Jay said. "Do you want a job?"

"Doing what?"

"News analyst for NBC."

Sam readily agreed. Then, realizing what he'd done, he added, "I've got to tell Mike, too." Mike Lafferty was his son-in-law and a reporter for *Florida Today*.

"I understand," Jay said. "But only after I go on the air at six o'clock. He can't have it in the paper before tomorrow morning anyway."

Sam hesitantly agreed.

Next, Jay came to see me to try to get a second source to confirm what he had learned—good reporters don't rely on just one source.

He came into my office in the press dome, closed the door, and sat down. "Have you heard what caused the accident?" he asked.

"No," I replied. "What are you hearing?"

"The O-ring seals on the right SRB failed. The flames burned through the casing and caused the explosion."

"I haven't heard that," I said. "Let me make some calls and try to confirm it. But it's your story. If HQ hasn't already planned a press conference and not told me, we won't ruin your scoop."

Jay didn't have to make any more phone calls. When he walked out of the dome, he saw Center Director Dick Smith about to drive off. "Mr. Smith, I just heard the cause of the accident was a burn through of the SRB. Can you confirm that?"

Surprised at how fast Jay had the story, he said, "You've got it."

Jay had another solid source.

That night at six-thirty p.m., Jay broke the biggest story of his life. He appeared on Tom Brokaw's evening news show, live by satellite from the press site, while other reporters listening nearby and then ran for their phones. Howard Benedict, Cape Canaveral bureau chief for the Associated Press, didn't have to listen. He and Jay had worked together many years, sharing information and helping each other fill in the blanks. Howard was ready to speed it through the wires as soon as Jay finished his report.

I had let NASA headquarters and KSC management know the story would be on the air. The next day we showed the press the camera view of the plume of fire streaming from the right SRB immediately before the entire vehicle was engulfed in flames.

CHAPTER NINE

Whose Fault Was It?

Soon after the accident, there were rumors of a meeting that took place the night before the launch that should have stopped it but didn't. Just like the press, the public affairs office hears hundreds of rumors and has to determine whether they're accurate before reacting in one way or another.

This one turned out to be true.

More than thirty Thiokol engineers, managers, and even vice presidents discussed concerns about the O-rings and cold temperatures with another half dozen or more officials from the Marshall Space Flight Center the day before launch.

These same concerns had been the subject of memos and warnings for years, but somehow were not discussed in the launch readiness reviews for *Challenger*. The Marshall solid-motor managers had gotten used to seeing hot gases get past the primary O-rings on the field joints and were trusting the secondary O-rings to prevent a breach.

In addition, dozens of engineers and managers who had been involved in examining motors after they flew had seen the blowby of hot gases and soot past the primary O-ring and some erosion of the secondary O-ring.

Yet, despite having all of the right people looking at the concerns, the information never reached final decision makers like Launch Director Gene Thomas and Associate Administrator Jesse Moore.

If I had any doubts about the value of having a presidential commission investigate the *Challenger* disaster, the hearings put those doubts to rest.

Without the commission, the story would have come out eventually in the course of the investigation. But written reports don't have the drama of courtroom-style testimony, or make as great an impression on people, whether within the program or in the general public. Nor, I think, would the launch decision process, the design concerns, and the technical details of the hardware have ended up being taught in almost every engineering school in the world had the information been presented differently.

The press played an important role as well. Once the cause of the accident was known, reporters scrambled to see how long the O-ring problem had been known. The *New York Times* and *Washington Post* printed articles revealing that even a budget analyst, Richard Cook, had mentioned O-ring concerns, pointing out the impact of an accident on future budgets. They also reprinted memos from Thiokol SRB Engineer Roger Boisjoly, in which he warned management about the danger.

The articles caught Commission Chairman Rogers's eye. The first day of hearings began with lengthy tutorials on the workings of every element of the space shuttle. Members learned about

NASA's multilayered management system for complex programs like the shuttle and Apollo before it. The Kennedy Space Center and its contractors were responsible for launch processing and launch. The Marshall Space Flight Center in Huntsville, Alabama, and its contractors were responsible for the development and manufacture of the external tank, the engines for the orbiters, and the solid rocket motors. The Johnson Space Center was responsible for the orbiter, astronaut training, and management of the mission once the vehicle cleared the tower.

Near the end of the first day, Rogers elicited the information he was really looking for when Marshall's Larry Mulloy introduced Allan McDonald into the conversation. McDonald was Morton Thiokol's solid rocket motor director, and he revealed the opposition to the launch by the Thiokol SRB engineers. It didn't come out in one cohesive story, but with McDonald's help the Commission was finally on the right path.

Shortly after noon on January 27, the team learned that temperatures of twenty degrees Fahrenheit were forecast for launch morning. Marshall's Solid Rocket Motor (SRM) Manager Larry Wear called Robert Ebeling, a counterpart at Morton Thiokol in Wasatch County, Utah, where the SRMs are manufactured. Wear asked Ebeling to have his engineers review the effect of very cold weather on the performance of the motors.

The twelve-foot-diameter, twenty-foot-long solid motor segments were transported from Utah to the Kennedy Space Center by railway car and then stacked together to form the complete motor. The segments fit together similarly to tongue-in-groove flooring, except that the tongues were thick steel and pinned together with 180 one-inch diameter steel pins around the circumference at each joint. Since these segments were assembled away from the factory, they were called field joints.

The Thiokol engineers' consensus was that the cold would affect the ability of the rubberlike Viton O-rings to seal the field joints in the solid motors, thereby allowing the six-thousand-degree gases to escape from the joints. They recommended that the launch be delayed until at least noon to allow the temperature to rise.

In order to include more of the SRM engineers and managers, another teleconference was arranged for eight fifteen p.m. This time, thirty-four engineers and managers participated, including high-level Thiokol and NASA/Marshall Space Flight Center managers.

Thiokol Vice President Joe Kilminster started the conference by stating that Thiokol could not recommend a launch when the temperature was "below fifty-three degrees Fahrenheit."

NASA SRB manager Mulloy bridled, reminding everyone that no launch-commit criteria had ever been established based on temperature. "My God, Thiokol, when do you want me to launch, next April?" he concluded.

NASA Deputy Director of Science and Engineering George Hardy added that he was "appalled" at the Thiokol recommendation.

Rather than continue, Kilminster asked for, and was granted, a five-minute recess to discuss the situation off-line with the Thiokol engineers and managers.

During the break, two engineers, Roger Boisjoly and Arnie Thompson, used sketches and photographs to explain the effect of cold on the O-rings. Boisjoly had fought hard to get a new design for the SRB joints. His memos detailing the problem and the length of time it had been known to Thiokol and NASA/Marshall managers were important pieces of the puzzle for the Commission and the press.

The O-rings had been damaged slightly on twelve of the twenty-four shuttle flights prior to *Challenger*. This had allowed hot gases to penetrate the primary O-ring, but the secondary seals had always held. A task force had been established to look into the problems the previous year but was not given enough funding or priority to proceed quickly.

One impediment to making a clear-cut decision was that the O-ring erosion had occurred when the weather was warm as well as cold.

Finally Thiokol Senior Vice President Jerry Mason said, "We have to make a management decision."

Three out of the four Thiokol managers voted to go ahead with the launch. Engineer Robert Lund still supported the engineering position. Mason told him, "Take off your engineering hat and put on your management hat." Lund finally agreed to vote with the others.

When the teleconference was resumed, Kilminster told the NASA managers that Thiokol had reversed its no-go recommendation and was ready to proceed to a launch. Thiokol was asked to put the recommendation in writing. The teleconference ended about eleven fifteen p.m.

Allan McDonald, Thiokol's SRM director, was already at KSC for the launch and not involved in the off-line deliberations. He told the NASA/Marshall managers at KSC, "If anything happens to this launch, I don't want to be the person that has to stand in front of a board of inquiry to explain why we launched outside the qualifications of the solid rocket motor or any shuttle system." He listed three reasons why the launch should be postponed: O-ring problems, booster recovery ships heading into wind toward shore due to high seas, and icing conditions on the launch pad.

McDonald was told in the meeting with Mulloy and others that "those are not your concerns," but that his comments would be passed on in an advisory capacity. McDonald's advice was not passed on, and the countdown continued.

CHAPTER TEN

The Search at Sea

Even before the Presidential Commission began its meticulous investigation into the launch, a huge fleet was combing through centuries of debris on the ocean floor.

NASA, aerospace contractors, and universities involved in space exploration are very good at diagnosing problems and failures even when a spacecraft is millions of miles away. If the *Challenger* accident had happened in Earth's orbit, I am sure they would still have accurately uncovered the causes. However, there is nothing like physical evidence to shorten an investigation and put it to rest for once and all.

That's where the armada of ships, aircraft, submarines, remotely operated submersibles, and men with names like Anker Rasmussen, Black Bart, and Ed O'Connor came in.

Rasmussen was in charge of the SRB recovery ships *Liberty Star* and *Freedom Star*. "Black Bart" was Captain Charles A. Bartholomew, the navy's superintendent of salvage. And Air Force

Colonel Edward O'Connor served as NASA's liaison with the search ships.

O'Connor originally arrived at Cape Canaveral Air Force Station in the early 1960s after volunteering to work on the Minuteman missile project. His first job was to "supervise the assembly of all the components for the Minuteman." He laughs and adds, "As much as a new second lieutenant *could* supervise experienced contractors."

His first interaction with NASA was when KSC's first director, Dr. Kurt Debus, chewed him out for destroying a launch pad. Failure was a frequent learning experience in the early days of rocket development.

O'Connor spent time in telemetry, overseeing ARIA tracking aircraft, and then was detailed to KSC. The air force was intimately involved with the development and use of the space shuttle. O'Connor watched the launch from a corner office in the headquarters building. As soon as the accident occurred, his supervisor simply told him, "Go on over and see what you can do to help."

Recalling the accident, O'Connor says, "When you have any sort of disaster, there is a period of time when it seems like chaos. It's a period when you are learning to deal with the issue emotionally, and figure out what needs to be done intellectually. For something as unexpected as *Challenger*, it took a while for all of that to come together.

"There was a lot of blame being passed around there. It wasn't blame for blame's sake; it was just confused reactions. What I had learned from being a launch director for several missions is, you have to put all that away and slow down to basic issues. You have to get done what has to be done."

Within an hour after the disaster, air and sea search-and-rescue teams were dispatched to the debris impact area approximately twenty miles offshore. Twelve aircraft from the air force, navy, and coast guard took part. During the first twenty-four hours, over one thousand square miles of the ocean were examined. Six hundred pounds of debris was brought in to port, most of it aluminum-covered honeycomb from the external tank. The largest piece retrieved the first day was fifteen by fifteen feet.

"I had come over as an observer and to do anything I could," O'Connor says. "But what slowly evolved was that I was taking care of recovery activities."

By January 31, he was working with NASA headquarters Manager of Aviation Safety Bill Comer, as well as representatives from the Air Force Safety and Investigation Center, the Army's Safety Center, and the various search groups.

The first job, O'Connor says, was consolidating all the bits and pieces. He established an office to serve as a single reporting point to the growing group.

"Our first analysis was that it would take at least a year to recover *Challenger* and cost about twenty-eight million dollars," O'Connor says. "But suddenly there was a presidential commission that was supposed to come up with a complete analysis of the problem within 120 days."

The original estimate was based on the navy using their normal search pattern: drawing a circle around the search area, dividing it into a grid, and slowly working inward from the perimeter.

O'Connor suggested starting with a trajectory analysis and energy analysis. "We have good tracking data," he told the search team. "However, that doesn't identify which piece is which. But if we go out in the debris field we can determine how the vehicle came apart. When we find one piece and it was on the opposite side of the vehicle from what we really want, then we should find what we're looking for 180 degrees from there."

A meeting with the radar experts allowed them to come up with an improved grid pattern that would yield results very quickly.

The Harbor Branch Oceanographic Institute was brought in. "They were the first underwater exploration organization to use GPS tracking. It was a big help," O'Connor says. On February 19, their submersible found the portion of the right solid motor where the burn through occurred.

"We had a lot of people who volunteered their expertise," O'Connor recalls. "Good people like Jacques Cousteau. We appreciated all of them, except it would have been just too chaotic to bring in any more groups. It was heartening to us to know how many people cared."

By the end of six days, over eleven tons of material had been recovered. The entire space shuttle, minus the propellants, weighed approximately 370 tons. Twenty-eight ships had searched over thirty-five thousand square nautical miles of ocean. Thirteen aircraft had flown 260 hours, searching sixty thousand square nautical miles.

Recovered debris was unloaded at the Trident Basin docks and trucked to one of three locations based on its type. Kenneth Colley, KSC's structures branch chief for vehicle engineering, was responsible for the placement of material in the Logistics Facility. He says, "As material was brought in, it was catalogued and placed on a grid pattern marked out on the floor or a similar area of the parking lot. The idea was to place them exactly in the position they would have been on the *Challenger* orbiter or the external tank."

Challenger wreckage retrieved from the Atlantic Ocean by a flotilla of U.S. Coast Guard and Navy vessels being offloaded at the Trident Basin at Cape Canaveral Air Force from the USCG Cutter Dallas, January 30, 1986. The black tile indicates it is part of the shuttle's fuselage.

Part numbers on most pieces pinpointed their original locations. "Large wooden supports were built," Colley recalls. "That way the sides could be placed in the right relationship with the bottom of the orbiter."

I escorted newspeople and photographers over to see the debris many times. To me, the layout resembled the area in a museum where pottery shards or dinosaur bones are assembled prior to putting them on display. But in the case of *Challenger*, the space had to be "super" dinosaur sized. The *Challenger* orbiter was the size of a commercial jet plane, and the external tank and solid rocket motors were taller than the Statue of Liberty.

The search for *Challenger* and its seven-member crew was the single largest search-and-recovery operation undertaken by the navy since the cleanup of harbors after World War II. The navy part of the operation involved more than five hundred divers, crew members, and support personnel, as well as ten surface ships and five submersibles to cover a 485-square-mile search area. The numbers were pushed far higher by the NASA, coast guard, air force, and private company ships and submersibles.

There were many days when the ships had to stay in port because of high seas, and many more that were not safe for divers. When divers could go down, their visibility was frequently about arm's length because sediment had been stirred up. Nevertheless, the divers and submersibles were vital to identifying objects and attaching lifting cables.

The highest priorities were assigned to finding and recovering the crew cabin and crew; the aft field joint portion of the right solid motor where the flames had been seen; the left solid motor; the left orbiter wing; and the tracking and data relay satellite with its upper stage motor.

CHAPTER ELEVEN

Putting Together the Pieces

Strong, decisive leaders are essential. But Bob Crippen and Jay Honeycutt also brought invaluable organizational and team-building skills to the group headed by Admiral Truly.

Crippen had been vice chairman of the NASA investigation from the beginning. He provided continuity among the leaders and the various committees. Within the first few days there were about six hundred people directly involved. It required spending most of his time in meetings with the various groups.

Chuck Hollinshead had been named public affairs representative for the NASA portion of the investigation. Widely respected across the agency, he was able to keep everyone in public affairs on the same page as far as what was known, thereby keeping the media from leaping to unwarranted conclusions.

When he was in town, Crippen met with Hollinshead or me every morning to review what was happening with the news media. He worked with each of the dozen other committees on an almost-daily basis to provide coordination and guidance. Crippen could be counted on to make quick decisions or find the right answer in a hurry. It was essential to making progress.

Honeycutt, who had worked with both Crippen and Truly in the past, helped bring some needed order in Washington.

"When the Presidential Commission was appointed by President Reagan," Honeycutt recalls, "everyone was given a set of NASA phone books. Each Commission member had a number of other people working for [him or her]. They all needed information and would look through the phone books, find branch chiefs and section heads that sounded like they were the right person, and call them with questions."

The directors of the various NASA centers grew concerned about the number of requests and whether the right people were being tasked with answering them. They asked Admiral Truly for help.

Truly designated Honeycutt to be NASA's liaison with the Commission, directing him to sit down with Executive Director Dr. Alton Keel and figure out how to give Commission members what they wanted.

"After the meeting," Honeycutt says, "I called George Abbey and asked to borrow Astronaut Bryan O'Connor. Next we set up an office at NASA headquarters. Then if one of the commissioners wanted something, they wrote it down and gave it to Keel. He gave it to Bryan, who gave it a number, helped determine who should have the action, and tracked it. Every day we were able to report how many open actions we had, who was doing what, and when the Commission could expect their answer.

"Bryan did a great job," says Honeycutt. "It completely settled the water between NASA and the Commission. I don't recall one complaint."

Honeycutt spent a lot of time going back and forth between Washington and KSC. He followed the salvage activities and watched over special projects. Although the *Challenger* mission had been unclassified, there were some classified electronic boxes that the Department of Defense was anxious to retrieve. Honeycutt had to report on the progress finding them every day until they were pulled from the water.

Although the Commission hearings, which started February 6, were of great import to the press and brought a lot of information to the table, the undersea search probably attracted more interest. One of the reasons was that very little information was emerging.

The navy and coast guard public affairs people did their best to brief the media on what was happening. The problem was that it was a very slow process.

Perhaps the biggest challenge was the lack of information about the crew and crew cabin. Finding the cabin and casting light on the last moments of national heroes would be big news. NASA wanted to protect the privacy and feelings of the families, but there simply wasn't anything to report yet.

Convinced that NASA had declared a news blackout, the media did their best to work around it.

One of the young, electronically savvy reporters, William Harwood of United Press International (UPI), led the charge.

Harwood rigged up a high-frequency radio to a tape recorder to capture the chatter between the search ships and between the ships and shore beginning at five a.m. every day and continuing through the entire work period. That way he could go back and listen to everything that happened whenever he came to work.

Other media organizations soon made use of the UPI antenna. They were convinced that NASA, the navy, and the coast guard knew they were listening and therefore had some "code words," which would come into play when they found the crew and crew cabin. Some guessed the code would be "T.J. O'Malley," after the tough, no-nonsense launch director who had launched John Glenn. Others thought "rudder" and "speedbrake" were the code words.

All of the NASA people who should have known if there was a code still swear that none existed. That group includes Ed O'Connor, who would be the one to drive the crew remains to Patrick Air Force Base when they were recovered.

Meanwhile, CBS—with better funding—joined in the game. They had access to night vision cameras that allowed them to discern what was on any search ship that came in at night. They stationed themselves at Cape Canaveral's Jetty Park to watch the ships come and go.

"Actually, I'm not sure it was worth it," UPI's Harwood says today. What did pay off was his dogged determination. He worked almost every day, and every day for more than eight hours. As a result, he was the only reporter there on the Sunday I was typing up a news release on how the crew cabin had finally been located.

On March 8, two divers, Terry Bailey and Mike McAllister, were almost out of breathing air and ready to go back to their ship for the day. The area they were in was smooth white sand, at a depth of ninety feet. Visibility was poor. Through the murky water they saw what looked like either another diver or an astronaut in a space suit. They knew the crew did not wear spacesuits during launch and realized it was the spacesuit that was on board *Challenger* in case an emergency spacewalk was needed.

They slowly moved forward and encountered the crumpled frame of the *Challenger* crew cabin—and within its twisted metal posts, the crew.

The best they could do with their remaining air supply was mark the find with a buoy and return to the ship.

During the next two days, the scene was documented, the bodies recovered, and the cabin lifted to the surface.

As I hovered over my word processor, Harwood said, "You look busy. What's happening?"

"We found the crew cabin," I told him.

"How much time do I have before you tell everyone else?"

"At least five minutes." Actually, he had more than that. Since no other reporters were there, I had to go through a long callout list in order to notify them.

Bill had his story on the wire to every country in the world long before most reporters knew it had happened.

The greatest value of the sea recovery program was to add what is called "ground truth" in other fields.

The video, the photographs, and the telemetry signals from *Challenger* provided electronic evidence of what had happened during the accident. The actual pieces, burned and twisted though they might be, showed beyond a shadow of doubt the truth of what had happened.

The recovered parts confirmed the failure scenario suggested by the photographs and ruled out a number of other possibilities.

CHAPTER TWELVE

Commission Conclusions

As the underwater search began to wind down, the Presidential Commission hearings were also coming to an end.

One cannot help but feel some sympathy for those who testified before the body. In addition to Chairman Rogers, who was a former secretary of state and U.S. attorney general, their questioners included a Nobel Prize laureate and the first person to set foot on another heavenly body.

The majority of those who testified were engineers. Some tried to give a tutorial on how everything worked, others overloaded their testimony with facts. All of them used acronyms. At one point the commissioners characterized their lengthy explanations as "sounding like a filibuster."

For example, Marshall Space Flight Center Solid Rocket Booster Manager Larry Mulloy testified for more than an hour about the assembly of the boosters, with particular emphasis on the seals and joints. He concluded by describing the complex tests being planned to understand how the joints behaved under the conditions prevailing during the launch of *Challenger*.

Following a ten-minute recess, Chairman Rogers announced that Physicist Richard Feynman wanted to respond to the presentation.

"Here is a comment for Mr. Mulloy," Dr. Feynman began. Holding up a Styrofoam cup of ice water, he lifted out a piece of O-ring that he had compressed with a C-clamp. He told Mulloy and the audience, "I took this stuff out of your seal and I put it in ice water, and I discovered that when you put some pressure on it for a while and then undo it, it doesn't stretch back. It stays the same dimension. In other words, for a few seconds at least, and more seconds than that, there is no resilience in this particular material when it is at a temperature of thirty-two degrees. I believe that has some significance for our problem." As previously stated, the temperature on launch day had been less than twenty degrees.

Dr. Feynman had just demonstrated that it didn't take a lot of complex tests to see why the *Challenger* launch should have been postponed.

Neil Armstrong probably asked fewer questions during the hearings than any of the other commissioners. However, since he was the primary person in charge of putting together the final report, one can conclude that his ideas formed a major part of it.

Years later, he told me that he saw his job in the hearings as gathering information, not expressing opinions. "The final report would do that." However, Armstrong's questions were always on target.

During testimony from Thiokol Vice President Robert Lund, Armstrong asked, "One question. Clearly, you had a concern about temperature, and so on. Did you ever consider or take thought of controlling the temperature at the seals, or to changing the material of the seals to something that had different characteristics?"

Lund acknowledged, "Those thoughts have gone through our minds. There has been no positive action along those lines."

During the redesign that was to follow, that's exactly what was done. Heaters were installed on each solid motor joint to ensure it never fell below the proper temperature again.

With the help of the testimony, NASA's own investigation, the contractors, and the evidence harvested from the ocean, the Commission concluded that the accident "was caused by the failure in the joints between the two lower segments of the right Solid Rocket Motor." They wrote, "The specific failure was the destruction of the seals that are intended to prevent the hot gases from leaking through the joint during the propellant burn of the rocket motor. The evidence assembled by the Commission indicates that no other element of the Space Shuttle system contributed to this failure." They added, "The failure was due to a faulty design unacceptably sensitive to a number of factors." Those factors included temperature.

What's more, the Commission stated, both Thiokol and the NASA solid rocket booster project office at Marshall failed to respond to facts obtained during testing, nor did they respond adequately to internal warnings about the faulty seal design.

Even more critical was this conclusion: "NASA and Thiokol accepted escalating risk apparently because they 'got away with it last time.'"

Commission member Feynman observed in the same section of the report that the decision-making process had been "a kind of Russian roulette. [The shuttle] flies [with O-ring erosion] and nothing happens. Then it is suggested, therefore, that the risk is no longer so high for the next flights. We can lower our standards a little bit because we got away with it last time. You got away with it, but it shouldn't be done over and over again like that."

In their four-month investigation, the Commission had examined every aspect of the shuttle, from the main engines to the liquid oxygen vent valve in the tip of the external tank, and found they had worked properly. The Commission/NASA team also reviewed every aspect of the prelaunch activities, procedures, and launch decision, along with organizational relationships and interactions.

Their final recommendations touched on almost every aspect of management and operations at NASA and its contractors. The list began with requiring redesign and replacement of the solid motor joint and seal, with independent oversight.

Greater safety underlined most of the recommendations, ranging from improved tires, brakes, and nosewheel steering to a new shuttle management structure that was more accountable to a headquarters manager than a center director. The Commission suggested an independent safety organization reporting to NASA headquarters and the use of more astronauts in safety and other management positions. The new associate administrator for safety would have the authority to stop a launch.

Marshall Space Flight Center was singled out to "energetically" eliminate a policy of "management isolation" that inhibited the flow of information about problems to NASA headquarters and elsewhere.

A complete review of parts and systems in which a single-point failure could result in loss of the vehicle or life was recommended.

Finally, the Commission suggested another look at methods for the crew to escape in an emergency. The space shuttle was the first and only American spacecraft that did not have a crew escape mechanism.

The commission also expressed strong support for the nation's space program. Their report read: "The Commission urges that NASA continue to receive the support of the Administration and the nation. The agency constitutes a national resource that plays a critical role in space exploration and development. It also provides a symbol of national pride and technological leadership."

In his transmittal letter accompanying the report to the president, Chairman Rogers stated, "Each member of the Commission shared the pain and anguish the nation felt at the loss of seven brave Americans in the *Challenger* accident on January 28, 1986. The nation's task now is to move ahead to return to safe space flight and to its recognized position of leadership in space. There could be no more fitting tribute to the *Challenger* crew than to do so."

CHAPTER THIRTEEN

The Crew

The Presidential Commission left the many questions about the last minutes of the crew to a NASA investigation headed by former Skylab Astronaut and Medical Doctor Joseph P. Kerwin. Kerwin submitted his report to Admiral Truly in mid-July, slightly more than a month after the Presidential Commission report.

>RADM Richard H. Truly
>Associate Administrator for Space Flight
>NASA Headquarters
>Code M
>Washington, DC 20546

>Dear Admiral Truly:
>   The search for wreckage of the *Challenger* crew cabin has been completed. A team of engineers and scientists has analyzed the wreckage and all other available evidence

in an attempt to determine the cause of death of the *Challenger* crew. This letter is to report to you on the results of this effort. The findings are inconclusive. The impact of the crew compartment with the ocean surface was so violent that evidence of damage occurring in the seconds [that] followed the explosion was masked. Our final conclusions are:

- the cause of death of the *Challenger* astronauts cannot be positively determined;
- the forces to which the crew were exposed during Orbiter breakup were probably not sufficient to cause death or serious injury; and
- the crew possibly, but not certainly, lost consciousness in the seconds following Orbiter breakup due to in-flight loss of crew module pressure.

Our inspection and analyses revealed certain facts which support the above conclusions, and these are related below: The forces on the Orbiter at breakup were probably too low to cause death or serious injury to the crew but were sufficient to separate the crew compartment from the forward fuselage, cargo bay, nose cone, and forward reaction control compartment. The forces applied to the Orbiter to cause such destruction clearly exceed its design limits. The data available to estimate the magnitude and direction of these forces included ground photographs and measurements from onboard accelerometers, which were lost two-tenths of a second after vehicle breakup.

Two independent assessments of these data produced very similar estimates. The largest acceleration pulse occurred as the Orbiter forward fuselage separated and was rapidly pushed away from the external tank. It then pitched nose-down and was decelerated rapidly by aerodynamic forces. There are uncertainties in our analysis; the actual breakup is not visible on photographs because the Orbiter was hidden by the gaseous cloud surrounding the external tank. The range of most probable maximum accelerations is from 12 to 20 G's in the vertical axis. These accelerations were quite brief. In two seconds, they were below four G's; in less than ten seconds, the crew compartment was essentially in free fall. Medical analysis indicates that these accelerations are survivable, and that the probability of major injury to crew members is low.

After vehicle breakup, the crew compartment continued its upward trajectory, peaking at an altitude of 65,000 feet approximately 25 seconds after breakup. It then descended striking the ocean surface about two minutes and forty-five seconds after breakup at a velocity of about 207 miles per hour. The forces imposed by this impact approximated 200 G's, far in excess of the structural limits of the crew compartment or crew survivability levels.

The separation of the crew compartment deprived the crew of Orbiter-supplied oxygen, except for a few seconds supply in the lines. Each crew member's helmet was also connected to a personal egress air pack (PEAP) containing an emergency supply of breathing air (not oxygen) for ground egress emergencies, which must be manually activated to be available. Four PEAP's were recovered, and there is evidence that three had been activated. The nonactivated PEAP was identified as the Commander's, one of the others as the Pilot's, and the remaining ones could not be associated with any crew member. The evidence indicates that the PEAP's were not activated due to water impact.

It is possible, but not certain, that the crew lost consciousness due to an in-flight loss of crew module pressure. Data to support this is:

- The accident happened at 48,000 feet, and the crew cabin was at that altitude or higher for almost a minute. At that altitude, without an oxygen supply, loss of cabin pressure would have caused rapid loss of consciousness and it would not have been regained before water impact.
- PEAP activation could have been an instinctive response to unexpected loss of cabin pressure.
- If a leak developed in the crew compartment as a result of structural damage during or after breakup (even if the PEAP's had been activated), the breathing air available would not have prevented rapid loss of consciousness.
- The crew seats and restraint harnesses showed patterns of failure which demonstrates that all the seats were in place and occupied at water impact with all harnesses locked. This would likely be the case had rapid loss of consciousness occurred, but it does not constitute proof.

Much of our effort was expended attempting to determine whether a loss of cabin pressure occurred. We examined the wreckage carefully, including the crew module attach points to the fuselage, the crew seats, the pressure shell, the flight deck and middeck floors, and feedthroughs for electrical and plumbing connections. The windows were examined and fragments of glass analyzed chemically and microscopically. Some items of equipment stowed in lockers showed damage that might have occurred due to decompression; we experimentally decompressed similar items without conclusive results.

Impact damage to the windows was so extreme that the presence or absence of in-flight breakage could not be determined. The estimated breakup forces would not in themselves have broken the windows. A broken window due to flying debris remains a possibility; there was a piece of debris imbedded in the frame between two of the forward windows. We could not positively identify the origin of the debris or establish whether the event occurred in flight or at water impact. The same statement is true of the other crew compartment structure. Impact damage was so severe that no positive evidence for or against in-flight pressure loss could be found.

Finally, the skilled and dedicated efforts of the team from the Armed Forces Institute of Pathology, and their expert consultants, could not determine whether in-flight lack of oxygen occurred, nor could they determine the cause of death.

/signed/

Joseph P. Kerwin

Astronaut Robert Overmyer, who was assigned to the NASA investigation, told me that both Commander Scobee and Pilot Smith had stayed awake long enough to try to fly the disembodied *Challenger* out of the inferno. Mike Smith had thrown several electrical switches and Scobee had tried to control the descent. Overmyer and Bob Crippen agreed that there was little chance they had stayed conscious through the entire descent. Their air packs held only sea-level pressure air and were designed for escaping the shuttle while still on the pad. But they were experienced space fliers who reacted with knowledge gathered over thousands of hours of training for a multitude of emergencies.

After examination, the crew remains were flown to Dover Air Force Base. Judy Resnik, Dick Scobee, and Mike Smith were buried individually by their families at Arlington National Cemetery. Ellison Onizuka was buried at the National Memorial Cemetery of the Pacific in Honolulu, Hawaii. Christa McAuliffe, Greg Jarvis, and Ronald McNair were buried together at the Space Shuttle *Challenger* Memorial at Arlington.

CHAPTER FOURTEEN

The Response

NASA and its contractors had not been sitting around and waiting for the Presidential Commission's recommendations. As problems surfaced in the course of the internal NASA investigation, work got underway to correct the deficiencies and plan for the future.

KSC reviewed every aspect of processing the shuttle for flight after the components and payloads arrived at the center. Those reviews yielded a fresh look at improving processes, equipment, and facilities. Even though nothing surfaced that contributed to the accident, many improvements suggested themselves.

One significant change was in the way overtime work impacted individual workers. *Challenger* Launch Director Gene Thomas, who was subsequently named KSC director of safety and quality assurance, says, "As a launch approached, the pressure for getting everything done to support a particular date got more and more intense." Managers like him had been pleased with records from the civil servant and contractor groups showing about a 20 percent rate of overtime. After all, that amounted to someone working six days as opposed to five. However, "the figures were calculated based on the total work force in that group," he told me. "Closer scrutiny revealed a few people working sixty or more hours a week. Some with special skills worked twelve hours or more in a day."

The rules were quickly changed.

The biggest changes in the Safety, Reliability, and Quality Assurance Directorate came from making it directly responsible to the administrator and center directors rather than to an adjunct for each discipline area. The associate administrator of safety now had the authority to stop a launch from proceeding. This was a major change resulting from the suggestions in the Presidential Commission report. In the past, most safety recommendations had been filtered through the various directors.

Lee Solid, site manager for Rocketdyne, the company that built the shuttle main engines, had worked on virtually every major rocket engine since the beginning of human space flight. During the *Challenger* accident he had been in the firing room, carefully monitoring the Space Shuttle Main Engines (SSME). The SSMEs were the most powerful engines per pound ever built. They used liquid oxygen and liquid hydrogen high-pressure turbo pumps that turned at approximately thirty thousand to thirty-five thousand revolutions per minute. Immediately after the accident, and even weeks later, people had speculated that they were the weak link in the system and predicted they were at fault.

As he followed the first minute of the flight through the firing room windows, Solid says, "I watched what was happening with a sense of disbelief. It was incredible. I didn't want to believe my eyes. What in the world could we have done [wrong]? There was nothing on our screens. Then after fifteen or twenty minutes, an eternity at the time, a Rocketdyne engineer came to the door. He had been talking to the plant and gave us a signal that the engines had worked perfectly . . . based on pump discharge pressures."

Still, Solid continued, there was room for improvement. The turbo pumps often needed refurbishing after a flight. A rail system had been developed to allow access for removing and replacing them.

"Over the years we had learned it was important to analyze our successes as well as our failures. In that analysis, we realized that it would save time, money—and improve safety—if we removed all three entire main engines after each landing and replace them with refurbished ones.

"During the time between *Challenger* and the next flight we had time and resources to build a dedicated engine shop at KSC, to make that easier to do," Solid says.

Improvements to the main engines were just a small part of the scores of modifications made before shuttle flights resumed.

The thermal-protection system, wings, tanks, fuel cells, and other parts and subsystems were all improved. Major work was done on the landing gears and tires.

All of the critical items and systems were reviewed and revised. The number of "Crit-1" items—single pieces of hardware or software whose failure would destroy the shuttle—doubled. So did time spent on inspection and testing.

The biggest job, obviously, would be the Solid Rocket Motors (SRM). Thiokol and the SRM group at MSFC went to work immediately.

Allan McDonald and engineer Roger Boisjoly were in the forefront of the redesign effort, just as they had been the primary voices warning of the problems before the accident. However, that assignment might never have happened.

In his 2009 book, *Truth, Lies, and O-Rings*, McDonald—the Thiokol manager who revealed his engineers' opposition to the *Challenger* launch—recounts how on March 3, 1986, he assigned all of his managers to tasks supporting the investigation and redesign of the motors. The next day, his boss's boss notified him that his assignments had been rescinded and that he was being transferred to another position removed from any direct responsibility for the solid motor redesign effort.

Thiokol could argue they had not fired him; he had the same salary as before and an important-sounding, if amorphous, title.

But after a number of changes in Thiokol management, McDonald was restored to his job of leading the redesign of the SRM joints a few months later. At the time, he said, he thought it was because they considered him the best person for the job. He learned later that Congressman Edward Markey (D-Mass.) had written to the CEO of Morton Thiokol, promising to bar the company from NASA contracts if it retaliated against either McDonald or Roger Boisjoly for their testimony to Congress or the Presidential Commission.

Boisjoly was particularly affected by the pressures of being considered a whistleblower by co-workers and neighbors according to McDonald. After some time off to rest, he left the company.

McDonald and his engineering team continued to plow ahead through the redesign process and qualification testing of the new hardware for flight.

Independent review boards were added both at the company and within NASA. While working with so many outside experts was time consuming, it also added additional expertise to the mix. Astronauts become more involved in every aspect of the technical work. Astronaut Robert (Hoot) Gibson was an ally in gaining approval for the new joint design. He would fly three more times after the resumption of shuttle flights.

The first half of 1986 was a tempestuous time for managers and directors who had been in charge before *Challenger*.

On May 12, in a White House ceremony, Dr. James C. Fletcher was sworn in as NASA Administrator for the second time. He had served as administrator from 1971 to 1977 during the inception of the space shuttle program and was an effective advocate during that time.

Former Astronaut Admiral Richard Truly took over the reins as associate administrator for space flight.

The center directors at Marshall Space Flight Center and Kennedy Space Center retired.

Jesse Moore, who had held the title of associate administrator of space flight and director of the Johnson space center, was moved back to NASA headquarters as NASA deputy general manager. Moore subsequently returned to Ball Aerospace in Boulder, Colorado, where he continued a distinguished career in advancing space technology.

Dozens of other changes were made in the management structures at every center. One major implementation at KSC was to isolate the job of launch director from the administrative duties formerly included in that position. Bob Sieck moved into that position and retired years later after having launched more people into space by the end of the shuttle program than any other launch director.

The Commission deemed communication about technical concerns a major problem. Within months new pathways for people to communicate with top management, even anonymously, were created.

A new crew escape mechanism was added. It required the shuttle orbiter to be in a stable position, but it did make it possible for the crew to get out safely if a crash was imminent.

Dozens of modifications were made to the thermal-protection system, wing, tanks, fuel cells, and other parts and subsystems of the orbiters. The tires and landing gear came in for major improvement as well.

The launch schedule pressure was relieved by flying Department of Defense missions only rarely and accepting no more commercial payloads. The move increased the use of expendable launch vehicles, but reduced the receipt of money from other sources to fund the shuttle program.

A high point within the changes was the appointment of Lieutenant General Forrest McCartney as KSC's center director.

When the announcement was made, the first reaction at KSC was not joy, but concern. Was this a move to militarize a proudly civilian agency?

It turned out he was the perfect man for that moment in history.

McCartney had grown up in the space business, rising from a young lieutenant in the position of air force satellite controller to commander of the highly classified air force space division. He understood what it was like to sit at a control console day after day; he also knew how to command thousands of people charged with keeping America safe. He was a physicist and a leader.

The employees at KSC were not the only ones who had been concerned at his new assignment. Years after he retired, McCartney told me he arrived at KSC with uncertainty and concern about the unknown. "I had spent my entire career dealing with the military and security, and I was going to a world that might not really want me."

Probably because it was my field, he went on to tell me, "It dawned on me to talk to the press, and I found them kinder and more considerate than I would have imagined. I was able to establish a rapport . . . and I just operated by never telling them anything I didn't want to see in print or on TV." Even better for us, he was willing to come out to the press site and talk to whoever was there in an informal way.

General McCartney at first found the work force "dissatisfied and demoralized." "But they were wonderful, spirited and highly motivated," he continued.

Almost every day he would drive out to a different work area, wander in, and ask people how they were doing. "I talked to technicians . . . they loved the program as much or more than anyone else," he said.

His own job, he confided, was, "One, keep their spirits up; two, be fair; and three, get them what they need to do their job."

Often, he would grab one of the other managers to go with him. He'd say, "Let's go kick some tires." He really meant, "Let's talk to the real workers." His credo was: "You know you're in the right job when you can whistle when you come to work and whistle when you go home."

One of the most telling things about the kind of man he was, and why he would have such a great effect on everyone at the center, was his Christmas ritual. Most everyone had the day off, but he made it a point to drive out to the Center, visit each guard post to say he appreciated them being there on the holiday, and give them a candy cane and cookies.

He helped KSC workers regain their pride. Finally, they were ready for the "Return to Flight": STS-26.

CHAPTER FIFTEEN

The Return to Flight

The time was 12:40 a.m., the date July 4, 1988. A really big rocket ship called the space shuttle slowly lumbered out of the vehicle assembly building and proceeded majestically toward launch pad 39 B. It was illuminated by the huge xenon lights that would spotlight it on launch day that coming September.

A crowd of thousands roared their approval. Most of them had spent the past two and half years working toward this moment. The orbiter *Discovery*, its external tank, and the solid rocket boosters had all undergone a rebirth.

As the shuttle passed a temporary stage set up for the occasion, a broadly smiling Center Director Forrest McCartney presented to Astronaut David Hilmers a book signed by 15,240 workers. The book would fly into space on STS-26 with the first American crew to blast off the Earth since the *Challenger* accident in January 1986.

In a sense it was a pledge that the thousands of people across America who had worked toward this launch had done their absolute best to ensure a safe journey for all who sailed on it.

The five-man crew would include Mission Specialist Hilmers, Commander Frederick Hauck, Pilot Richard Covey, and Mission Specialists John M. Lounge and George D. Nelson. It was the first time a space shuttle crew would be made up entirely of veteran astronauts.

Hilmers told the crowd, "It is those of you who have written your names in this book who have made this splendid, magnificent sight we behold tonight possible. But tonight you haven't given this book and this shuttle to just the five members of our crew. Indeed you have given it to all Americans. It is the mark of a great nation . . . that it can rise again from adversity, and with *Discovery*, rise again we will."

The journey from VAB to launch pad took the usual six-plus hours. And in the morning, before dawn broke, more than four thousand cars carrying workers and their families were lined up outside the gates for a chance to drive around the pad to see the shuttle before the protective rotating service tower closed around it.

In keeping with recommendations from the Presidential Commission, Hauck and his crew had been invited to sit in on all reviews throughout the recovery period. As they traveled from center to center across the U.S., they spoke with engineers and technicians in places like the orbiter processing facility, the engine shop, and the vehicle assembly building.

The astronauts had always believed that forming relationships with the people their lives depended on was only common sense. Most enjoyed getting to know the men and women, many of whom had devoted their working lives to the space program.

Earlier in the shuttle program, the astronauts had developed the Space Flight Awareness Program to honor workers who had made exceptional contributions. An astronaut presented him or her with a certificate and Snoopy pin designed by cartoonist Charles Schulz, and the award included front-row seats in the VIP stands for a launch.

The fruit of these visits often showed up in surprising ways. Months before rollout, the crew was invited to KSC for a ceremony, during which *Discovery* would be powered up for the first time after it had been overhauled. Hauck says, "Imagine having this most complex machine in the world, and someone takes the power cord and plugs it into the wall. Energy surges through it, and you're hoping that you're not going to smell smoke and everything works."

There was no smoke.

The orbiter workers were all in brightly colored Hawaiian shirts. General McCartney next presented the astronauts with their own Hawaiian shirts, explaining that the facility workers called themselves the "Loud and Proud" group. The crew were now honorary members.

After he thanked everyone for the work they were doing, Hauck promised them, "We're going to take these shirts up into space with us."

Launch morning—September 29, 1988—was warm and pleasant, in sharp contrast to the last time I had driven in for a launch. Even at three in the morning, the causeways were filling up with cars. Eventually, there would be about a million viewers, as many or more than had attended the very first shuttle launches.

Because of safety concerns, far fewer people were invited to the center than for any other launch. Hollinshead had to fight long and hard for every single person. The media had been restricted to no more than five hundred at the press site. Everyone had had to prove that they could not do their job unless they were there. More than 1,500 other reporters were relegated to the causeway, four miles away. They were not happy.

The firing room was calm and businesslike when I arrived shortly after the astronauts had breakfast. A high percentage of the people there had experienced *Challenger*. They were confident and determined that this would be the smoothest countdown and launch ever.

Things did go smoothly, with two minor exceptions. The rules had changed during the long look at safety after *Challenger*, and the crews were now required to wear bright orange partial pressure suits. The launch was delayed an hour and a half to replace fuses in the cooling system of two of the crew's suits.

Another constraint was waived after careful debate about upper atmospheric wind shear.

Discovery roared to life at 11:37:00 a.m. eastern standard time, and while all of us held our breath, the shuttle climbed steadily skyward.

We all breathed again as the solid motors finished their burn and dropped away. The crew's first job, after settling into orbit and wriggling out of their space suits, was launching a tracking and data relay satellite, thus completing the primary mission of *Challenger*. Hauck had joked during an interview that they had gone through training for this mission so many times, they could do everything "with their eyes closed." The deployment went perfectly.

Each of the crew took time to say a few words about the crew of *Challenger*. It was difficult, it was heartfelt, and it was necessary for moving on.

It's been a tradition from the beginning of the shuttle program for mission control to wake the crew with a song that is special to one of its members. Usually the crew has no idea what the wake-up call will be, although their families and co-workers have all been polled for ideas.

After the crew's first sleep period, comedian Robin Williams provided "wake-up call" levity when he reprised his famous line from the movie *Good Morning, Vietnam*. This time, he said, "Good Morning, *Discovery*!" and mentioned each of the crew members by name.

Space flight is stressful but it can also be fun. It's important to keep a balance. A few days later it was time to don those "loud and proud" Hawaiian shirts as they flew over Hawaii.

America had returned to space, but not without a warning for the future. After the crew members landed and were greeted by President George H. W. Bush, it was learned that *Discovery* had suffered severe damage to its thermal-protection tiles in the underwing area. Post-flight analysis showed the cause was impact from a twelve-inch piece of cork insulation that had ripped loose during ascent.

Ironically, the cork tore from the forward field joint of the right-hand solid motor. Nothing on the shuttle had received more scrutiny than the solid motors. And yet it happened.

The culprit in the 2003 *Columbia* disaster was also a light-weight piece of insulation that came loose and impacted the orbiter. But a very small mass hitting at the speed of sound or more can do deadly damage.

Flying in space will probably never be without danger. Yet, since primitive humans first looked at the heavens, the stars have beckoned. Humans have always responded to the best of their ability.

CHAPTER SIXTEEN

The Challenge Remains

"The *Challenger* was lost—but not the challenge." That was the late-night message a ninety-two-year-old grandmother expressed to one press site worker shortly after the accident. The call was one of thousands from all over the world expressing sorrow along with a fervent desire to see the space program move forward. Along with the calls came tens of thousands of letters from people of all ages, on every continent.

What the grandmother urged so eloquently was a renewal of the spirit that had inspired people to advance knowledge throughout history.

I hope she is pleased with what she saw resulting from the accident. Hopefully, she lived to see the shuttle fly again when she was ninety-five, and became aware of the myriad of programs that resulted from the many people inspired by the accident to work even harder.

Ordinary people often see the value the space program brings to the country better than politicians, who try to balance short-term benefits from direct aid to constituents versus long-term benefits to everyone. Those values are never clearer than when a problem like the *Challenger* accident occurs and is overcome.

Continuing the space shuttle program and the international space station was a real bonanza, one that will go on for years.

There have always been at least three vital reasons for a strong space program. The first acknowledges that technology is the most important resource the United States has to sell to the rest of the world. The second is advancing our own and the world's capability to live and work in space and on this planet. The third is showing the rest of the world the advantages of using resources for peaceful expansion of human capability.

Advances in technology come from working on and learning from solving difficult problems. Going to the moon was once thought impossible by many, and building a large functioning national laboratory in space was, too.

NASA has always worked hand in hand with American industry to manufacture the technology and products needed by the space program. New knowledge acquired in the process can be used immediately to develop consumer products and other purposes. This arrangement helped strengthen industries from aircraft to robotics and helped them compete more effectively with the rest of the world.

The shuttle gave us the ability to reach up to ailing spacecraft, grab them by hand—or by robotic arm—repair them, and redeploy or bring them back to the Earth. The Hubble space telescope is the best example of servicing on orbit. Hubble never would have had such a prolific career if that had not been possible.

A total of 105 satellites were launched during the shuttle program's thirty years of service.

Putting America's scientists to work on orbit has contributed to every discipline from medicine to meteorology.

Technology from the main shuttle engines led to the development of a miniaturized ventricular assist pump, a joint project by NASA and famous heart surgeon Dr. Michael DeBakey. The tiny device is just two inches long by one inch in diameter.

A form of space anemia in astronauts led to the development of new, easier ways to diagnose and treat anemia on earth. Foam used to insulate the shuttle's external tank was repurposed to produce master molds for prosthetic limbs.

Hundreds of other technologies have spun off of the shuttle and space station program to become commercial products. Lighting technology developed for plant growth experiments on the space station now treats brain tumors in children. Infrared sensors developed to remotely measure the temperatures of distant stars are now available at the corner drug store to take the temperature of an adult or child instantly and painlessly.

Stretching our intellectual muscles by doing difficult things has changed the world we live in. There is practically no area that has not been improved by space technology. The expansion of world communication, long-term weather forecasting, and the knowledge of global warming by themselves are probably worth the entire investment.

Little things count, too. Jay Honeycutt, who became center director of Kennedy Space Center ten years after the *Challenger* accident, spearheaded the transfer of the silicon protective tile technology on the space shuttle to NASCAR racecars. A few ounces of the tile material prevented race drivers' feet from being burned. That had been a problem for decades.

From the inception of the astronaut corps, promoting education has been a personal passion of the astronauts. The original astronauts drew on that passion to form the Mercury 7 Foundation to raise money and award scholarships to deserving young students. Now known as the Astronaut Scholarship Foundation, the group presents almost thirty $10,000 scholarships each year with a goal of awarding one scholarship per year for each of the eight-seven astronauts in the Astronaut Hall of Fame. To date, $3.7 million has been disbursed to top students across the country. Many of them are now working in research and development in dozens of fields.

The families of the *Challenger* crew, under strong leadership from Commander Scobee's wife, June, created an educational foundation—the Challenger Center for Space Science Education—that has advanced and inspired hundreds of thousands of young people. The first Challenger Learning Center was opened in 1988; it has grown to a network of nearly fifty centers around the globe.

The core of each learning center is an interactive computerized simulator with a mission control room patterned after the one at the NASA Johnson Space Center, as well as a digital orbiting space station ready for exploration. Each year more than four hundred thousand students participate in space-themed missions that have been developed to strengthen their knowledge and interest in science, technology, and mathematics.

The Astronaut Memorial Foundation, headquartered at the Kennedy Space Center, also grew out of the *Challenger* tragedy. It has taken on a dual role: honoring all astronauts who have lost their lives in the pursuit of space exploration, and the education of America's young people. It prepares teachers to use the latest knowledge and technology in their classrooms across the country.

The training, as judged by participant evaluations, is among the best in the nation. More than 9,100 technology specialists, teachers, administrators, educational decision makers, and university professors have attended AMF training. The foundation estimates that following attendance in an AMF program, each participant has a direct impact on approximately 120 students. Using this multiplier effect, AMF programs have touched over ten million students since 1995.

Dozens of schools across the country have been named either for *Challenger* or one of the individual astronauts. Even a drawbridge on Florida's Merritt Island was named for Christa McAuliffe.

Over the years, the lessons from the *Challenger* accident have become an integral part of the engineering curriculum in colleges and universities. They are discussed in classes not just covering the nuts and bolts of engineering but also its ethics.

The American Association for the Advancement of Science pointedly recognized the importance of ethics in engineering in a recent award. Its Prize for Scientific Freedom and Responsibility was awarded to Thiokol engineer Roger Boisjoly "for his honesty and integrity leading up to and directly following the shuttle disaster."

The real tragedy of an event like *Challenger* is in the loss of people and the accomplishments and inspiration they would have contributed to humankind. One can only guess what talented people like the crew of *Challenger* might have done had they returned to Earth alive.

Yet, even their loss has inspired thousands of others to carry on for them in meaningful ways we will never fully know. They have not been lost to the world.

They made it a better place.

ACKNOWLEDGMENTS

In addition to the people interviewed for the book, the memories and files of the following storytellers were vital:

Robert (Bob) Murray
Great photos from helicopters

Carol Cavanaugh
Secretary and guide to directors

Bud Crisafulli
Grove owner and land developer

George Diller
Public Affairs specialist and my successor as launch commentator

ACKNOWLEDGMENTS

George English
KSC director of management and advisor

Kay Grinter
Librarian and writer, KSC Press Site

Darleen Hunt
Writer, editor, protocol specialist, and fine artist

Elaine Liston
NASA KSC archivist and history expert

Thomas O'Toole
Washington Post space expert

Maggie Persinger
KSC's photographic envoy to thousands of writers

Jill Rock
A welcoming presence in KSC protocol

Klaus Wilckens
KSC's top photographer for decades

Thanks also to my editor at Open Road, Betsy Mitchell.

ABOUT THE AUTHOR

Called "the Voice of NASA" for many years by the world's television networks, Hugh Harris devoted thirty-five years with the National Aeronautics and Space Administration to telling the story of the United States space program. Although he is best known to the public for his calm, professional commentary on the progress of launch preparations and launch of the space shuttle, his primary accomplishments were in directing an outreach program to the general public, news media, students, and educators, as well as to business and government leaders. He also oversaw the largest major expansion (up to that time) in the history of the Kennedy Space Center's visitor complex and tours.

Harris began his career as a member of the news media. He worked as a reporter and broadcaster for WMTR in Morristown, New Jersey, and as a reporter and photographer for two newspapers.

After his retirement in 1998, he shared his experience in NASA public relations with nuclear industry leaders at conferences held by the United Nations' International Atomic Energy Agency in Europe and Japan and in this country through the Nuclear Energy Institute.

He continues to work as a volunteer at the KSC Press Site, as well as for the Astronaut Scholarship Foundation.

Find a full list of our authors and titles at www.openroadmedia.com

FOLLOW US
@OpenRoadMedia

CPSIA information can be obtained
at www.ICGtesting.com
Printed in the USA
JSHW021828220223
38099JS00003B/536